Charles Joyce White

The elements of theoretical and descriptive astronomy for the use of colleges and academies

Charles Joyce White

The elements of theoretical and descriptive astronomy for the use of colleges and academies

ISBN/EAN: 9783337275761

Printed in Europe, USA, Canada, Australia, Japan

Cover: Foto ©berggeist007 / pixelio.de

More available books at **www.hansebooks.com**

THE ELEMENTS

OF

THEORETICAL AND DESCRIPTIVE

ASTRONOMY,

For the use of Colleges and Academies.

BY

CHARLES J. WHITE, A.M.,

ASSISTANT PROFESSOR OF MATHEMATICS IN HARVARD COLLEGE.

FOURTH EDITION, REVISED.

PHILADELPHIA:
CLAXTON, REMSEN & HAFFELFINGER.
1880.

Entered according to Act of Congress, in the year 1869, by
CHARLES J. WHITE,
in the Clerk's Office of the District Court of the United States for the District of Maryland.

PREFACE TO THE THIRD EDITION.

THE first edition of this work was published in 1869, to meet the requirements of the students of the United States Naval Academy. In preparing it, I endeavored to present the main facts and principles of Astronomy in a form adapted to the elementary course of instruction in that science which is commonly given at colleges and the higher grades of academies. I selected those topics which seemed to me to be the most important and the most interesting, and arranged them in the order which experience had led me to believe to be the best. The third edition of the book is issued with no change in the general plan, and with only those changes which the advance of astronomical knowledge in the last six years renders necessary.

In the descriptive portions of the work, I have endeavored to give the latest information upon every topic which is introduced. On not a few points the opinions of competent observers are by no means the same; and on these points I have endeavored to give, as far as possible, the various opinions which now exist. The distances and the dimensions of the heavenly bodies are given to correspond with the value of the solar parallax which is at present adopted in the American Ephemeris; the recent theories upon the connection of comets and meteors, the principles of spectroscopic observation, and the conclusions concerning the constitution and the movements of the heavenly bodies which such observation induces, are given, it is hoped, in sufficient detail.

No clear conception of the processes by which most of the fundamental truths of Astronomy have been established can be attained without some knowledge of Mathematics. I have endeavored, however, to confine the theoretic discussions within the limits of such moderate mathematical knowledge as may fairly be expected in those readers for whom the treatise is intended. Certain definitions and formulæ, with which the student may possibly not be familiar, will be found in the Appendix; and, with this aid, I believe that every portion of the work can be read without difficulty.

In the preparation of the work, many authorities have been consulted; the principal ones being Chauvenet's *Manual of Spherical and Practical Astronomy*, and Chambers's *Descriptive Astronomy*. The treatise has been used as a text-book in the United States Naval Academy, the Massachusetts Institute of Technology, Harvard College, and other institutions; and I am indebted to officers of these institutions for many valuable suggestions as to errors and improvements. I trust that this new edition will be found to be free from mistakes, and that it will be useful, not only to the class of students for whom it is especially prepared, but to others who may wish to know the general principles and the present state of the science of Astronomy.

Harvard College, Cambridge, Mass., 1875.

CONTENTS.

The Greek Alphabet...PAGE ix

CHAPTER I.
GENERAL PHENOMENA OF THE HEAVENS. DEFINITIONS. THE CELESTIAL SPHERE.

The heavenly bodies. Astronomy. Form of the earth. Diurnal motions of the heavenly bodies. Right, parallel, and oblique spheres. Definitions. Theorems. The astronomical triangle. Spherical coordinates. Vanishing lines and circles. Spherical projections....... 11

CHAPTER II.
ASTRONOMICAL INSTRUMENTS. ERRORS.

The clock: its error and rate. The chronograph. The transit instrument: its construction, adjustment, and use. The meridian circle. The reading microscope. Fixed points. The mural circle. The altitude and azimuth instrument. Method of equal altitudes. The equatorial. The sextant. The artificial horizon. The vernier. Other astronomical instruments. Classes of errors........................ 28

CHAPTER III.
REFRACTION. PARALLAX. DIP OF THE HORIZON.

General laws of refraction. Astronomical refraction. Geocentric and heliocentric parallax. The dip of the horizon............................ 51

CHAPTER IV.
THE EARTH. ITS SIZE, FORM, AND ROTATION.

Measurement of arcs of the meridian by triangulation. Spheroidal form of the earth. Its dimensions. Its volume, density, and weight. Rotation of the earth. Change of weight in different latitudes. Centrifugal force. The trade-winds. Foucault's pendulum experiment. Linear velocity of rotation............................ 59

CHAPTER V.

LATITUDE AND LONGITUDE.

Four methods of finding the latitude of a place. Latitude at sea. Reduction of the latitude. Longitude. Greenwich time by chronometers, celestial phenomena, and lunar distances. Difference of longitude by electric and star signals. Longitude at sea. Comparison of the local times of different meridians........................PAGE 72

CHAPTER VI.

THE SUN. THE EARTH'S ORBIT. THE SEASONS. TWILIGHT. THE ZODIACAL LIGHT.

The ecliptic. Distance of the sun from the earth determined by transits of Venus. Magnitude of the sun. The earth's orbit about the sun. The seasons. Twilight. Rotation of the sun, and its constitution. The zodiacal light.. 83

CHAPTER VII.

SIDEREAL AND SOLAR TIME. EQUATION OF TIME. THE CALENDAR.

The sidereal and the solar year. Relation of sidereal and solar time. The equation of the centre. The equation of time. Astronomical and civil time. The calendar.. 100

CHAPTER VIII.

UNIVERSAL GRAVITATION. PERTURBATIONS IN THE EARTH'S ORBIT. ABERRATION.

The law of universal gravitation. The mass of the sun. The earth's motion at perihelion and aphelion. Kepler's laws. Precession. Nutation. Change in the obliquity of the ecliptic. Advance of the line of apsides. Diurnal and annual aberration. Velocity of light. Aberration a proof of the earth's revolution.. 107

CHAPTER IX.

THE MOON.

The orbit of the moon, and perturbations in it. Variation of the moon's meridian zenith distance. Distance, size, and mass of the moon. Augmentation of the semi-diameter. The phases of the moon. Sidereal and synodical periods. Retardation of the moon. The harvest moon. Rotation; librations and other perturbations. The lunar cycle. General description of the moon........................ 120

CHAPTER X.

LUNAR AND SOLAR ECLIPSES. OCCULTATIONS.

Lunar eclipses. The earth's shadow. Lunar ecliptic limits. Solar eclipses. The moon's shadow. Solar ecliptic limits. Cycle and number of eclipses. Occultations. Longitude by solar eclipses and occultations..PAGE 135

CHAPTER XI.

THE TIDES.

Cause of the tides. Effect of the moon's change in declination. General laws. Influence of the sun. Priming and lagging of tides. The establishment of a port. Cotidal lines. Height of tides. Tides in bays, rivers, &c. Four daily tides. Other phenomena................... 146

CHAPTER XII.

THE PLANETS AND THE PLANETOIDS. THE NEBULAR HYPOTHESIS.

Apparent motions of the planets. Heliocentric parallax. Orbits of the planets. Inferior planets. Direct and retrograde motion. Stationary points. Evening and morning stars. Elements of a planet's orbit. Heliocentric longitude of the node. Inclination of the orbit. Periodic time. Mercury. Venus. Transits of Venus. Superior planets: their periodic times and distances. Mars. The minor planets. Bode's law. Jupiter: its belts, satellites, and mass. Saturn and its rings. Disappearance of the rings. Uranus. Neptune. The nebular hypothesis.. 155

CHAPTER XIII.

COMETS AND METEORIC BODIES.

General description of comets. The tail. Elements of a comet's orbit. Number of comets and their orbits. Periodic times. Motion in their orbits. Mass and density. Periodic comets. Encke's comet. Winnecke's or Pons's comet. Brorsen's comet. Biela's comet. D'Arrest's comet. Faye's comet. Méchain's comet. Halley's comet. Remarkable comets of the present century. The great comet of 1811. The great comet of 1843. Donati's comet. The great comet of 1861. Meteoric bodies. Shooting stars. The November showers. Height and orbits of the meteors. Detonating meteors. Aerolites. Connection of comets and meteoric bodies 187

CHAPTER XIV.

THE FIXED STARS. NEBULÆ. MOTION OF THE SOLAR SYSTEM. REAL MOTIONS OF THE STARS.

Proper motions of the fixed stars. Magnitudes. Constellations. Constitution of the stars. Distance of the stars. Bessel's differential observations. Real magnitudes of the stars. Variable and temporary stars. Double and binary stars. Colored stars. Clusters. Resolvable and irresolvable nebulæ. Annular, elliptic, spiral, and planetary nebulæ. Nebulous stars. Double nebulæ. The Magellanic clouds. Variation of brightness in nebulæ. The milky way. Number of the stars. Motion of the solar system in space. Real motions of stars detected with the spectroscope........................PAGE 213

APPENDIX.

Mathematical definitions, theorems, and formulæ.............................. 242
Kirkwood's law... 247
Chronological history of astronomy... 248
Sketch of the history of navigation... 255
Table I.—Elements of the planets, the sun, and the moon............ 256
" II.—The earth... 257
" III.—The moon... 257
" IV.—Elements of the satellites.. 258
" V.—The minor planets... 259
" VI.—Schwabe's observations of the solar spots.................... 261
" VII.—Elements of the periodic comets................................ 261
" VIII.—Transits of the inferior planets................................. 262
" IX.—Stars whose parallax has been determined................... 262
" X.—The constellations... 263
" XI.—Examples of variable stars... 266
" XII.—Examples of binary stars.. 267

INDEX 269

THE GREEK ALPHABET.

The following table of the small letters of this alphabet is given for the use of those readers who are unacquainted with the Greek language.

α	Alpha.	ν	Nu.
β	Bēta.	ξ	Xi.
γ	Gamma.	o	Omĭcron.
δ	Delta.	π	Pi.
ε	Epsĭlon.	ρ	Rho.
ζ	Zēta.	σ	Sigma.
η	Eta.	τ	Tau.
ϑ or θ	Thēta.	υ	Upsĭlon.
ι	Iōta.	ϕ	Phi.
κ	Kappa.	χ	Chi.
λ	Lambda.	ψ	Psi.
μ	Mu.	ω	Omēga.

ASTRONOMY.

CHAPTER I.

GENERAL PHENOMENA OF THE HEAVENS. DEFINITIONS.

1. *The heavenly bodies* are the sun, the planets, the satellites of the planets, the comets, the meteors, and the fixed stars.

The planets revolve about the sun in elliptical orbits, and the satellites revolve in similar orbits about the planets. The earth is a planet, as we shall see hereafter, and the moon is its satellite. The comets revolve about the sun in orbits which are either ellipses, parabolas, or hyperbolas. Comparatively little is known with any degree of certainty about the meteors; but it is probable that they too revolve about the sun.

The sun, the planets, the satellites, and the comets constitute what is called *the solar system*. The fixed stars are bodies which lie outside of this system, and preserve almost precisely the same configuration from year to year.

The heavenly bodies may be considered to be projected upon the concave surface of a sphere of indefinite radius, the eye of the observer being at the centre of the sphere. This sphere is called *the celestial sphere*.

2. *Astronomy* is the science which treats of the heavenly bodies. It may be divided into *Theoretical, Practical,* and *Descriptive Astronomy*.

Theoretical Astronomy may be divided into *Spherical* and *Physical Astronomy*.

Spherical Astronomy treats of the heavenly bodies when considered to be projected upon the surface of the celestial sphere. It embraces those problems which arise from the apparent diurnal motion of the heavenly bodies, and also those which arise from any changes in the apparent positions of these bodies upon the surface of the celestial sphere.

Physical Astronomy treats of the causes of the motions of the heavenly bodies, and of the laws by which these motions are governed.

Practical Astronomy treats of the construction, adjustment, and use of astronomical instruments.

Descriptive Astronomy includes a general description of the heavenly bodies; of their magnitudes, distances, motions, and configuration; of their appearance and structure; of, in short, every thing relating to these bodies which comes from observation or calculation.

3. *Form of the Earth.*—We may assume, at the outset, that the form of the earth is very nearly that of a sphere. The following are some of the reasons which may be given for such an assumption:—

(1.) If we stand upon the sea-shore, and watch a ship which is receding from the land, we shall find that the topmasts remain in sight after the hull has disappeared. If the surface of the sea were merely an extended plane, this would not happen; for the topmasts, being smaller in dimensions than the hull, would in that case disappear first. The supposition that the surface of the sea is curved, however, fully accounts for this phenomenon, as may be seen in Fig. 1. Let the curve CBG represent a portion of the earth's surface, and let A

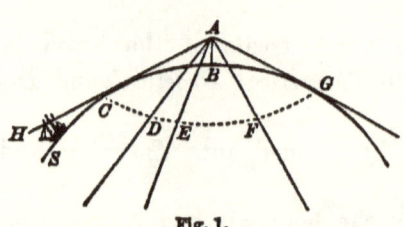

Fig. 1.

be the position of the observer's eye: it is at once evident that no portion of the ship, S, will be visible which is situated below the line AH, drawn from A tangent to the earth's surface at C. The same figure also shows why it is that when a ship is approaching land any object on shore can be seen from the topmasts before it is seen from the deck.

(2.) At sea, the visible horizon everywhere appears to be a circle. This also is easily explained on the supposition that the earth is spherical in form; for if, in Fig. 1, the line AH is turned about the point A, and is continually tangent to the sphere, the points of tangency, C, D, E, &c., will form the visible horizon of the spectator, and will evidently constitute a circle.

(3.) A lunar eclipse occurs when the earth is situated between the moon and the sun. Now the shadow which the earth at such a time casts upon the moon is invariably circular in form: and a body which in every position casts a circular shadow must be a sphere.

4. *Diurnal motion of the heavenly bodies.*—Two things will be noticed by an observer who watches the heavens during any clear night. The first is, that all the heavenly bodies, with the exception of the moon and the planets, retain constantly the same *relative* situation; and the second is, that all these bodies, without any exception whatever, are continually changing their positions with reference to the horizon. Let us suppose the observer to be at some place in the Northern Hemisphere. A plane passed tangent to the earth's surface at his feet will be his

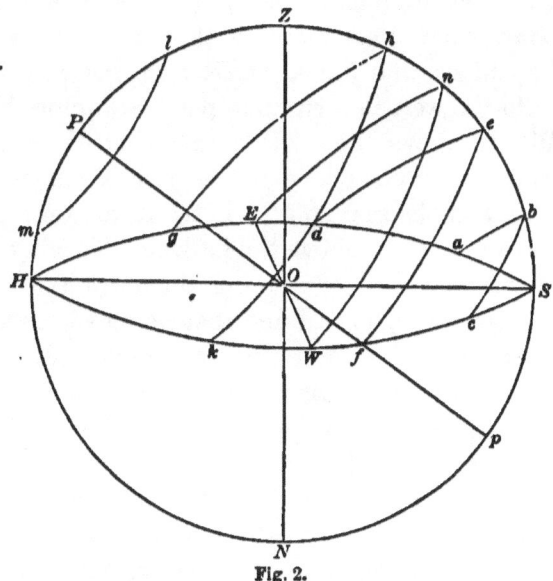

Fig. 2.

sensible horizon, and a second plane, parallel to this, passed through the centre of the earth, will be his *rational horizon*. If these two planes be indefinitely extended in every direction, they will intersect the surface of the celestial sphere in two circles; but the radius of the celestial sphere is so immense in comparison with the radius of the earth, that these two circles will sensibly coincide, and will form one great circle of the sphere, called the *celestial horizon*. In other words, the earth, when compared with the celestial sphere, is to be regarded as only a point at its centre.

Let Fig. 2, then, represent the celestial sphere, at the centre of which, O, the observer is stationed. Let the circle $HESW$ be his celestial horizon, of which H is the north point, S the south, E the east, and W the west.

If he looks towards the southern point of the horizon, and watches the movements of some star which rises a little east of south, he will see that it rises above the horizon in a circular path for a little distance, attains its greatest elevation above the horizon when it bears directly south, and then descends and passes below the horizon a little west of south. In the figure, *abc* represents the path of such a star. If he notices a star which rises more to the eastward, as at d, for instance, he will see that it also passes from the east quarter of the horizon to the west in a circular path, attaining, however, a greater altitude above the horizon than that which the star a attained, and remaining above the horizon a longer time. A star which rises in the east point will set in the west point, and will remain twelve hours above the horizon. Turning his attention to some star which rises between the north and the east, as the star g, for instance, he will find that its movements are similar to the movements of the stars already noticed, and that it will remain above the horizon for more than twelve hours. Finally, if he turns towards the north, he will see certain stars, called *circumpolar stars*, which never pass below the horizon, but continually revolve about a fixed point in the heavens, very near to which point is a bright star called the *Pole-star*, which, to the naked eye, appears to be stationary, though observation shows that it also revolves about this same fixed point. In the figure, *lm* represents the orbit of a circumpolar star.

If the same course of observations be repeated on the following night, the observer will find that the situations of the stars with reference both to each other and to the horizon are the same that they were when he first began to examine them; the motions which have already been described will be repeated, the circumpolar stars will still revolve about the same point in the heavens, and, in short, all the phenomena of which he took note will again be exhibited.

5. *Inferences.*—Three important truths are proved by a series of observations similar to that which we suppose to have been made.

(1.) The points at which the path of each star intersects the horizon remain unchanged from night to night, as long as the geographical position of the observer remains the same.

(2.) All the stars, whether they move in great or in small circles, make a complete revolution in identically the same interval of time;—that is to say, in twenty-four hours.

(3.) If we call that point about which the circumpolar stars appear to revolve the *north pole* of the heavens, and call the right line drawn from this point through the common centre of the earth and the celestial sphere the *axis* of the celestial sphere, the planes of the circles of all the stars are perpendicular to this axis.

Whether, then, the earth remains at rest, and the celestial sphere rotates about its axis, as above defined, or the celestial sphere remains at rest, and the earth rotates within it on an axis of its own, one thing is certain: the axis of rotation preserves in either case a constant direction. If the celestial sphere rotates about its axis, this axis always passes through the same points of the earth's surface; and if the earth rotates within the celestial sphere, its axis of rotation is constantly directed to the same points on the surface of the celestial sphere.

We shall see hereafter the reasons which have led to the adoption of the theory that these apparent motions of the stars are really due to the rotation of the earth upon its own axis. At present, for the sake of convenience in description, we shall consider the earth to be at rest, and shall speak of the apparent motions of the heavenly bodies as though they were real.

6. *Farther observations.*—Let us now suppose that the observer

leaves the place where he has hitherto been stationed, and travels in the direction of the point about which the circumpolar stars have appeared to revolve, and which we have called the north pole of the heavens. The general character of the phenomena which he observes will not be changed; but he will notice a change in this respect: the elevation of the north pole above the horizon will continually increase as he travels towards it, and the planes of the circles of the stars, remaining constantly perpendicular to the axis of the celestial sphere, will become less and less inclined to the plane of the horizon. The consequence of this will be that stars which are near the southern point of the horizon will remain a shorter time above the horizon, and will finally cease to appear; while in the northern quarter of the heavens the number of stars which never pass below the horizon will continually increase. Finally, if we suppose the observer to go on until the north pole is directly above his head, the stars which he sees will neither set nor rise, but will continually move about the sphere in circles whose planes are parallel to the plane of the horizon. In such a situation, it is evident that half of the celestial sphere will be perpetually invisible to him. Referring to Fig. 2, such a state of things is represented by supposing the line OP to be moved up into coincidence with OZ, the planes of the circles abc, def, &c., still remaining perpendicular to OP. The stars a and d will lie continually below the horizon, and the stars g and l continually above it.

Such a sphere as this just now described, where the planes of revolution are parallel to the plane of the horizon, is called a *parallel sphere*.

If the observer, instead of travelling towards the north pole, travels directly from it, its elevation above the horizon will continually decrease, and the obliquity of the planes of the circles to the plane of the horizon will continually increase, until he will at length reach a point at which the north pole will lie in his horizon and the planes of the circles will be perpendicular to its plane. Referring again to Fig. 2, the line PO will in this case coincide with OH, and the arcs abc, def, &c., will have their planes perpendicular to the plane of the horizon. It is evident that at this point every star in the celestial sphere will come above the horizon once in twenty-four hours, and that half of every circle will lie above the horizon, and half below it.

The geographical position which the observer has now reached is some point on the earth's *equator* (Art. 7). Such a sphere as this, where the planes of the circles are perpendicular to the plane of the horizon, is called a *right sphere*. Besides the right and the parallel sphere, we have also the *oblique sphere*, where the planes of the circles are oblique to the plane of the horizon. Such a sphere is represented in Fig. 2.

If the observer travels still farther in the same direction, the north pole will sink below his horizon, and the other extremity of the axis of the celestial sphere, called the *south pole*, will rise above it. There will be circumpolar stars revolving about this pole, and, in brief, all the phenomena which the observer noticed while travelling towards the north pole will be repeated as he travels towards the south pole.

DEFINITIONS.

7. We are now prepared to define certain points, angles, and circles on the earth and on the celestial sphere.

The *axis* of the celestial sphere is, as we have already seen, an imaginary line drawn from the north pole of the heavens through the common centre of the earth and of the celestial sphere, and produced until it again meets the surface of the celestial sphere in the south pole. The points where this axis meets the surface of the earth are called the *north pole* and the *south pole* of the earth.

That pole of the heavens which is above the horizon at any place is called the *elevated* pole at that place; the other is called the *depressed* pole.

The *axis* of the earth is that diameter which passes through the poles of the earth.

The earth's *equator* is a great circle of the earth, whose plane is perpendicular to the axis. This circle is, of course, equidistant from the two poles, and divides the earth into two *hemispheres*. That hemisphere which contains the north pole is called the *northern* hemisphere, the other is called the *southern* hemisphere.

Parallels of latitude are small circles of the earth whose planes are perpendicular to the axis.

Terrestrial meridians are great circles of the earth passing through the poles.

The *latitude* of any place on the earth's surface is its angular distance from the plane of the equator. This angle is measured by the arc of the meridian included between the place and the equator. Latitude is reckoned, either north or south, from 0° to 90°.

The *longitude* of any place is the inclination of its own meridian to the meridian of some fixed station, and is measured by the arc of the equator included between these two meridians. Longitude is usually reckoned east or west of the fixed meridian, from 0° to 180°. The meridian of Greenwich, England, is most commonly taken for the fixed meridian, though the meridians of Washington, Paris, and other places are also taken for the same purpose. The fixed meridian is called the *prime* meridian.

The arc of a parallel of latitude included between any two meridians is the *departure* between those meridians for that latitude. It is evident that if the departure between any two meridians is taken on two different parallels of latitude, the departure at the greater latitude will be the smaller.

Upward motion at any place is motion from the centre of the earth. *Downward* motion is towards the same point.

The *celestial horizon* has already been defined (Art. 4). Lines drawn perpendicular to the plane of the horizon are called *vertical* lines. The vertical line at any place, if indefinitely prolonged, meets the celestial sphere in two points. The upper point is called the *zenith*, the lower point the *nadir*.

The *celestial meridian* of any place is the great circle in which the plane of the terrestrial meridian of that place, when indefinitely produced, meets the surface of the celestial sphere. The axis of the sphere divides the celestial meridian into two semicircumferences: that which lies on the same side of the axis as the zenith is called the *upper branch* of the meridian, and the other is called the *lower branch*. The points where the celestial meridian and the celestial horizon intersect are called the *north* and the *south point* of the horizon,—the point which is the nearer to the north pole being the north point. The line in which the planes of these same two great circles intersect is called the *meridian line*.

Vertical circles are great circles of the sphere which pass through the zenith and nadir. That vertical circle the plane of which is perpendicular to the plane of the celestial meridian is called the *prime vertical*. The points in which the prime vertical cuts the horizon are called the *east* and the *west point* of the horizon; and the line which joins these two points is the *east and west line*.

The *celestial equator*, also called the *equinoctial*, is the great circle of the sphere in which the plane of the earth's equator, indefinitely produced, meets the celestial sphere.

The *altitude* of a heavenly body is its angular distance above the plane of the celestial horizon, measured on a vertical circle passing through that body. The *zenith distance* of the body is its angular distance from the zenith, and is evidently the complement of the altitude.

The *azimuth* of a celestial body is the inclination of the vertical circle which passes through the body to the celestial meridian, and is measured by the arc of the celestial horizon included between this vertical circle and the celestial meridian. Azimuth may be reckoned from either the north or the south point of the horizon, and towards either the west or the east. Navigators usually reckon it from the north point in north latitude and the south point in south latitude, and east or west as the body is east or west of the meridian, thus restricting it numerically to values less than 180°. Astronomers, on the contrary, usually reckon from the south point, to the right hand, from 0° to 360°.

The *amplitude* of a heavenly body is its angular distance from the prime vertical when in the horizon. It is reckoned from the east point when the body is rising, and from the west point when the body is setting, towards the north or the south as the body is to the north or the south of the prime vertical.

The *vernal equinox* is a certain fixed point upon the equinoctial. It is also called the *first point of Aries.*

Hour circles, or *circles of declination*, are great circles of the sphere passing through the poles of the heavens.

The *right ascension* of a heavenly body is the inclination of its hour circle to the hour circle which passes through the vernal equinox; or it is the arc of the equinoctial intercepted between these two hour circles. Right ascension is usually reckoned in hours, minutes, and seconds (an hour being taken equal to 15° of arc), and is *always* reckoned to the *eastward*, from 0h. to 24h.

The *declination* of a heavenly body is its angular distance from the plane of the celestial equator, measured on the circle of declination passing through the body. It is reckoned in degrees, minutes, &c., to the north and the south. The *polar distance* of a body is its angular distance from either pole, measured on its hour circle. Usually, however, when we speak of the polar distance of a body, we mean its angular distance from the elevated pole.

If, in Fig. 2, with the arc *HP*, which measures the altitude of the elevated pole, as a polar radius, we describe a circle about the pole as a centre, it is evident that the stars whose circles lie within this circle will never set. This circle is called the *circle of perpetual apparition.* It is equally evident that stars whose circles lie within a circle of the same magnitude, described about the depressed pole as a centre, will never come above the horizon. This circle is therefore called the *circle of perpetual occultation.*

The passage of a celestial body across the meridian is called its *transit* or *culmination.* When the body is within the circle of perpetual apparition, both transits occur above the horizon, one above the pole, the other below it. These are called the *upper* and the *lower* transit. For all bodies outside this circle, and not within the circle of perpetual occultation, the upper transit occurs above the horizon, the lower below it. For all bodies whatever, the upper transit occurs when the body crosses the upper branch of the meridian, and the lower transit when it crosses the lower branch.

The *hour angle* of a heavenly body is the inclination of the circle of declination which passes through the body to the celestial meridian, and is measured by the arc of the celestial equator included between these two circles. Hour angles are reckoned positively *towards the west*, from the upper culmination, from 0° to 360°, or 0h. to 24h.

The hour angle of the sun is called *solar time*, and that of the first point of Aries *sidereal time*. The interval of time between two consecutive upper transits of the sun is called a *solar day*, and the interval between the upper transits of the first point of Aries is called a *sidereal day*. The celestial sphere apparently makes one revolution about the earth in a sidereal day. The solar day is, on the average, about 3m. 56s. longer than the sidereal day.

8. Some of the preceding definitions are illustrated in the diagram, Fig. 3. In this figure O is the position of the observer, $HESW$ his celestial horizon, Pp the axis of the heavens, P the elevated and p the depressed pole, Z the zenith, and N the nadir. The circle $HZSN$ is the observer's celestial meridian. It may be noticed that this circle is at once a vertical and an

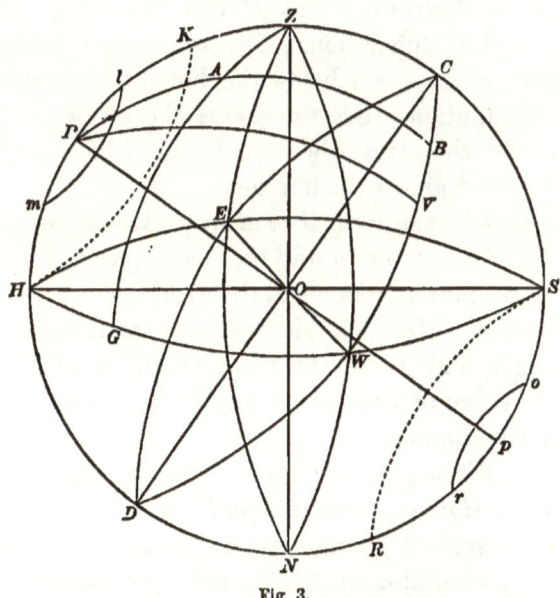

Fig. 3.

hour circle. The circle $ECWD$ is the equinoctial, and the circle $EZWN$, perpendicular to the meridian, is the prime vertical, cutting the horizon in E and W, the east and the west point of the horizon. The equinoctial, being also perpendicular to the meridian, passes through the same points. If P is supposed to be the north pole, H is the north point of the horizon, and S the south point.

Let A denote some celestial body. GA is its altitude, ZA its zenith distance, HG its azimuth as reckoned by navigators, and SG its azimuth as reckoned by astronomers, all these elements of position being determined by the arc of a vertical circle, ZG, passed through A. Let an arc of an hour circle, PB, be also passed through A. Then is AB its declination, PA its polar distance, and if V be taken to denote the position of the vernal equinox, VB is the right ascension of A. The right ascension may also be represented by the angle VPB, which the arc VB measures. The angle ZPB is the hour angle of A, and ZPV is the hour angle of the vernal equinox, or the sidereal time. This angle may also be designated as the *right ascension of the meridian*.

The circle KH, drawn about P with the radius PH, is the circle of perpetual apparition. The star whose path is represented by lm never passes below the horizon: l is its upper, m its lower culmination. SR represents the circle of perpetual occultation, and the stars whose paths lie, like or, within this circle, never come above the horizon.

9. *Theorem.*—*The sidereal time at any place is always equal to the sum of the right ascension and the hour angle of the same body.* This is an important astronomical theorem, and is readily proved in Fig. 3, in which ZPV, the sidereal time, is the sum of ZPB, the hour angle, and VPB, the right ascension of the celestial body A. Any two of these three angles, then, being given, the third is readily obtained.

Corollary.—*When a celestial body is at its upper culmination at any place, its right ascension is equal to the sidereal time at that place.* This is evidently true, because the hour angle of a body when at its upper culmination is zero, and hence, from the theorem, the right ascension of the body is equal to the sidereal time. For example, in Fig. 3, if l is a body at its upper culmination, VPZ is both its right ascension and the right ascension of the meridian.

10. *Theorem.*—*The latitude of any place on the earth's surface is equal to the altitude of the elevated pole at that place.* Let L (Fig. 4) be some place on the earth's surface, Pp the earth's axis, and EQ the equator. The line HR, tangent to the earth's surface at L, is the horizon, and Z the zenith, of L. According to the definition already given, LOQ is the latitude of L. Let the earth's axis be indefinitely prolonged, and at L let the line LP'', parallel to the earth's axis, be also indefinitely prolonged. Owing to the immensity of the celestial sphere when compared with the earth, these two lines will sensibly meet at a common point on the surface of the celestial sphere, and this common point will be the elevated pole. The elevated pole, then, to an observer at L, will lie in the direction LP'', and $P''LH$ will be its altitude.

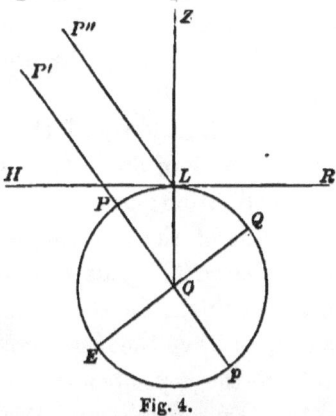

Fig. 4.

Now we have, $\quad HLZ = POQ$
$$ZLP'' = ZOP'$$
$$\therefore P''LH = LOQ$$
which was to be proved.

11. *Theorem.*—*The latitude of a place is equal to the declination of the zenith of that place.* This is easily seen in Fig. 4: for the declination of any body or point is its angular distance from the plane of the celestial equator, and hence ZOQ in the declination of the zenith.

Either of these theorems might be deduced from the other, but it is better to consider them as independent propositions.

12. *The Astronomical Triangle.*—The spherical triangle PZA (Fig. 3) is called the *astronomical triangle*. It is formed by the arcs of the meridian of the place, and of the vertical circle and the hour circle passing through some heavenly body, which are included between the zenith of the observer, the elevated pole, and the position of the body as projected on the surface of the celestial sphere. The three sides are: ZP, the co-latitude of the

place; PA, the polar distance of the body; and ZA, its zenith distance. The three angles are: ZPA, the hour angle of the body; PZA, its azimuth; and PAZ, an angle which is rarely used, and which is commonly called the *position angle* of the body.

The co-latitude and the zenith distance can evidently never be greater than 90°. The polar distance is equal to 90° *minus* the declination: and it is less than 90° when the body is on the same side of the celestial equator as the observer, but becomes *numerically* 90° *plus* the declination when the body is on the opposite side. In the former case the declination of the body is said to have the *same name* as the latitude, in the latter case, to have the *opposite name*.

13. *Diurnal Circles.*—We have already seen that the apparent daily motions of the stars are performed in circles, the planes of which are perpendicular to the axis of the sphere. These circles are called *diurnal circles*. The phenomena which have been observed with reference to these circles (Arts. 4 and 6) are explained in Fig. 3. The circles of all the stars which rise in the arc of the horizon, ES, are evidently divided by the horizon into two unequal parts, the smaller of which in each case lies above the horizon. Hence these stars are above the horizon less than twelve hours. On the other hand, the greater portions of the circles of those stars which rise between E and H lie above the horizon, and the stars themselves are above it for more than twelve hours.

Any body, then, whose declination is of the same name as the elevated pole will be above the horizon more than twelve hours, while a body whose declination is of a different name from that of the elevated pole will be above the horizon less than twelve hours. And further: those bodies which are *in north declination*, in other words, to the north of the celestial equator, will rise to the north of east and set to the north of west; while bodies in south declination will rise and set to the south of the east and the west point.

When a star has no declination, that is to say, is on the celestial equator, it will rise due east and set due west, and will remain twelve hours above the horizon.

14. *Right and Parallel Spheres.*—When an observer travels towards the elevated pole, the radius of the circle of perpetual apparition, being equal to the altitude of the pole, continually increases, and the number of stars which never set increases in like manner. The number of stars which never rise also increases. Finally, if he reaches the pole, the celestial equator coincides with the horizon, the east and the west point disappear, and the bodies which are on the same side of the equator with the observer are perpetually above the horizon, and revolve in circles whose planes are parallel to it, while the bodies which are on the opposite side of the equator never rise. As he travels towards the equator, the circles of perpetual apparition and occultation alike diminish, the diurnal circles become more and more nearly vertical, and when he reaches the equator, the equinoctial becomes perpendicular to the horizon and coincides with the prime vertical, and the horizon bisects all the diurnal circles. At the equator, then, every celestial body comes above the horizon, and remains above it twelve hours.

15. *Spherical Co-ordinates.*—The position of any point on the surface of a sphere is determined, as soon as its angular distances are given from any two great circles on that sphere whose positions are known. Thus the geographical position of any point on the earth's surface is known when we have determined its latitude and longitude; in other words, when we know its angular distance from the equator and from the prime meridian. In like manner we know the position of any point on the surface of the celestial sphere when either its altitude and azimuth, or its right ascension and declination, are given. In the first of these two systems of co-ordinates the fixed great circles are the celestial horizon and the celestial meridian, the origin of co-ordinates being either the north or the south point of the horizon. In the second system the fixed great circles are the equinoctial and the circle of declination which passes through the vernal equinox, and the origin of co-ordinates is the vernal equinox. In Fig. 3, if we know the arcs HG and GA, or the arcs VB and BA, we evidently know the position of A.

16. *Vanishing Points and Vanishing Circles.*—Every one knows that as he increases the distance between himself and any object,

the apparent magnitude of the object decreases; and that, if he recedes far enough from it, it will be reduced in appearance to a point. Every one also knows that when he looks along the line of a railroad track the lines appear to converge, and that, if the track is straight, and the curvature of the earth does not limit his vision, the rails will ultimately appear to meet.

These familiar illustrations will serve to show what is meant by a *vanishing point*. The actual distance between the rails of course remains the same; but the angle at the eye which this distance subtends decreases as the eye is directed along the track, until at last it ceases to subtend any appreciable angle at the eye, and the rails apparently meet. This point where the rails appear to meet is called the *vanishing point* of the two lines; and, in general, the vanishing point of any system of parallel lines is the point at which they will appear to meet, when indefinitely prolonged. We have already seen (Art. 10) that the pole of the heavens is the vanishing point of lines drawn perpendicular to the equator, and the same may be said of the poles of any circle on the celestial sphere. For instance, the poles of the horizon at any place are the zenith and the nadir; and any system of lines perpendicular to the horizon will apparently meet, when prolonged indefinitely, in these two points. And again, the east and the west point of the horizon are the poles of the meridian, and lines drawn perpendicular to the meridian will have these points for their vanishing points.

The same principle holds good when applied to any system of parallel planes. They will appear to meet, when indefinitely extended, in one great circle of the sphere, and this circle is called the *vanishing circle* of that system of planes. The celestial horizon is, as has already been stated (Art. 4), the vanishing circle of the planes of the sensible and the rational horizon. and indeed of any number of planes passed parallel to them. The celestial equator is the vanishing circle of the planes of all the parallels of latitude, and, in short, every circle of the celestial sphere may be regarded as the vanishing circle of a system of planes passed perpendicular to the line which joins the poles of that circle.

17. *Spherical Projections.*—The points and circles of either the earth or the celestial sphere, or of both, may be projected upon the plane of any great circle of either sphere. The plane on which the projections are made is called *the primitive plane*, and the circle which bounds this plane is called *the primitive circle*. Several distinct methods of projection will be found in treatises on Descriptive Geometry. Of these, the most common are the *orthographic*, the *stereographic*, and *Mercator's* projection.

In the *orthographic* projection, the point of sight is taken in the axis of the primitive circle, and at an infinite distance from that circle. All circles whose planes are perpendicular to the primitive plane are projected into right lines; all circles whose planes are parallel to the primitive plane are projected into circles, each of which is equal to the circle of which it is a projection; and all other circles are projected into ellipses.

In the *stereographic* projection, the point of sight is at either pole of the primitive circle, and its distance from that circle is finite. In this projection every circle is projected as a circle, unless its plane passes through the point of sight, in which case it is projected into a right line.

Mercator's projection is employed in the construction of charts representing the earth's projection. In this projection the parallels of latitude are represented by parallel right lines, and the meridians are also represented by parallel right lines, perpendicular to the equator. The meridian projections are equidistant, but the distance between the successive latitude projections increases as we recede from the equator. The advantage offered by this projection to navigators is that the ship's track, as long as the course on which it sails is unaltered, is represented on the chart by a straight line, and that the angle which this line makes with each meridian is the course.

CHAPTER II.

ASTRONOMICAL INSTRUMENTS.   ERRORS.

18. This chapter will be devoted to a general description of the common astronomical instruments, of the class of observations to which each is adapted, and of the manner in which such observations are made. No attempt will be made to describe the elaborate mechanism by which, in many cases, the usefulness of the instrument is increased and its manipulation is facilitated; but enough, it is hoped, will be said to enable the student to form a clear conception of the prominent features of each instrument which is described. There is no lack of excellent treatises on Astronomy, in which those who wish to investigate this subject more thoroughly will find all the details, which the limits prescribed to this book will not permit to enter here, clearly and elaborately presented.

THE ASTRONOMICAL CLOCK.

19. The astronomical clock is a clock which is regulated to keep sidereal time, and is an indispensable companion to the other astronomical instruments. It is provided with a pendulum so constructed that change of temperature will not affect its length. The sidereal day at any place commences, as has already been stated, when the vernal equinox is on the upper branch of the meridian of that place, and the theory of the sidereal clock is that it shows 0h. 0m. 0s. when the vernal equinox is so situated. Practically, however, it is found that every clock has a *daily rate;* that is to say, it gains or loses a certain amount of time daily. In order, then, that a clock may be regulated to sidereal time, it is necessary to know both its *error* and its *daily rate;* the *error* being the amount by which it is fast or slow at any given time, and the *daily rate* being the

amount which it gains or loses daily; and knowing these, it is evidently in our power to obtain at any desired instant the true sidereal time from the time shown by the face of the clock.

It is to be noticed further, that, except as a matter of convenience, a small rate has no advantage over a large one; but it is very important that the rate, whether large or small, shall be *constant* from day to day; so that, of two clocks, one of which has a large and constant rate, and the other a small and varying one, the preference is to be given to the former.

Clocks may be regulated to keep either local sidereal time or Greenwich sidereal time, or both.

20. *Error of the Clock.*—To obtain the error of a clock on the local sidereal time at any observatory, we make use of the proposition, already demonstrated (Art. 9), that the right ascension of any celestial body, when at its upper culmination on any meridian, is equal to the sidereal time at that meridian. The Nautical Almanac gives the right ascensions of more than a hundred stars which are suitable for observations for time. By means of an instrument, properly adjusted, we determine the instant when any one of these stars is on the meridian, and the time which the clock shows at that instant is noted. This is the *clock time of transit*, and the right ascension of the star, taken from the Almanac, is the *true time of transit;* and a comparison of these two times will evidently give us the amount by which the clock is fast or slow on local sidereal time.

21. *Daily Rate.*—If, in a similar manner, we obtain the error of the clock on the next or on any subsequent day, the difference of these two errors will be the gain or loss of the clock in the interval; and hence, if we divide this difference by the number of the days and parts of days which have intervened between the two observations, the quotient will be the daily gain or loss.

22. *Chronograph.*—The accuracy of astronomical observations is much enhanced by recording the times of the observations by means of an electric current. A cylinder, about which a roll of paper is wound, is turned about on its axis with a uniform motion by the use of appropriate machinery. A pen is pressed

THE TRANSIT INSTRUMENT.

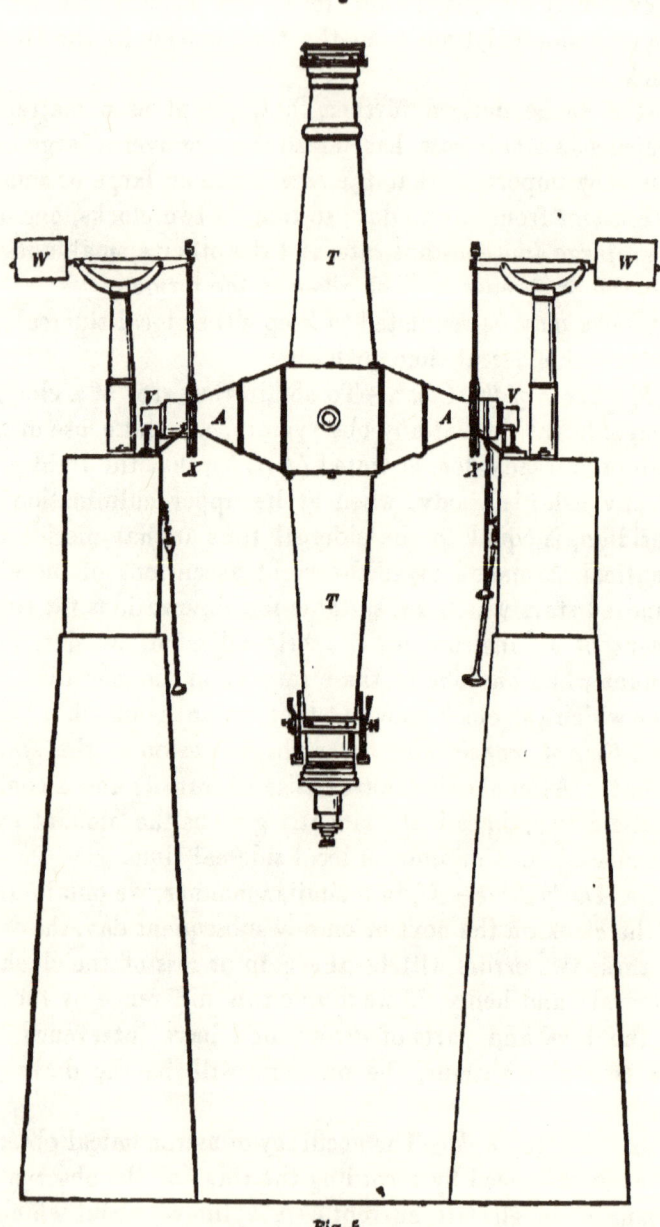

Fig. 5.

down upon the paper, and is so connected with a battery that whenever the circuit is broken a mark of some kind is made upon the paper. The wires of this battery are connected with the sidereal clock in such a way that every oscillation of the pendulum breaks the circuit. Every second is thus recorded upon the revolving paper. The observer also holds in his hands a *break-circuit key*, with which, whenever he wishes to note the time, he breaks the circuit, and thus causes the pen to make its mark upon the paper. The line which the pen describes upon the paper will be something like this:

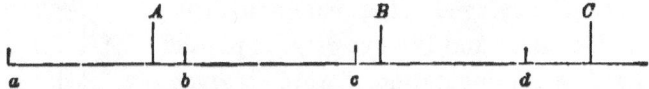

The equidistant marks, *a, b, c,* &c., are the marks caused by the pendulum, and the marks *A, B, C,* are the marks which the pen makes when the circuit is broken by the observer. The distance from *A* to *b, B* to *c,* &c., can be measured by a scale of equal parts, and the time of an observation can thus be obtained within a small fraction of a second.

The cylinder is also moved by a fine screw in the direction of its own length, so that the pen records in a spiral.

This instrument is called a *Chronograph.* There are several varieties of the chronograph in use by different astronomers, but the main principle in all of them is similar to that of the instrument just now described.

THE TRANSIT INSTRUMENT.

23. The transit instrument is used, as its name implies, in observing the transits of the heavenly bodies.. Fig. 5 represents a transit instrument. It consists of a telescope, *TT*, sustained by an axis, *AA*, at right angles to it. The extremities of this axis terminate in cylindrical pivots, which rest in metallic supports, *VV*, shaped like the upper part of the letter Y, and hence called the *Ys*. These Ys are imbedded in two stone pillars. In order to relieve the pivots of the friction to which the weight of the telescope subjects them, and to facilitate the motion of the

telescope, there are two counterpoises, *WW*, connected with levers, and acting at *XX*, where there are friction rollers upon which the axis turns. When the instrument is properly adjusted, the telescope, as it turns about with the axis *AA*, will continually lie in the plane of the meridian; and, in order to effect this, the axis *AA* should point to the east and the west point of the horizon, and be parallel to its plane. There are therefore screws at the ends of the axis, by which one extremity of the axis may be raised or depressed, and may also be moved forward or backward.

24. *The Reticule.*—In the common focus of the object-glass and the eye-glass is placed the *reticule*, a representation of which is given in Fig. 6. It consists of several equidistant vertical wires (usually seven) and two horizontal ones. If the instrument is accurately adjusted to the plane of the meridian, the

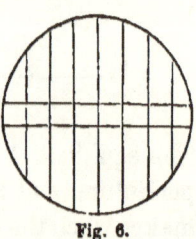

Fig. 6.

instant that any star is on the middle wire is the instant of its transit. These wires are also called the *cross-wires*.

25. *Adjustment.*—The *axis of rotation* of the instrument is an imaginary line connecting the central points of the pivots.

The *axis of collimation* is an imaginary line drawn from the optical centre of the object-glass, perpendicular to the axis of rotation.

The *line of sight* is an imaginary line drawn from the optical centre of the object-glass to the middle wire.

The transit instrument is accurately adjusted in the plane of the meridian, when the line of sight of the telescope lies continually in that plane, as the telescope revolves. Three things, then, are readily seen to be necessary: the axis of rotation must be exactly horizontal; it must lie exactly east and west; and the line of sight and the axis of collimation must exactly coincide. Practically, these conditions are rarely fulfilled; but they can, by repeated experiments, be very nearly fulfilled, and the errors which the failure rigorously to adjust the instrument causes in the observations will be constant and small, and can be accurately determined.

26. *Application.*—The principal application of the transit

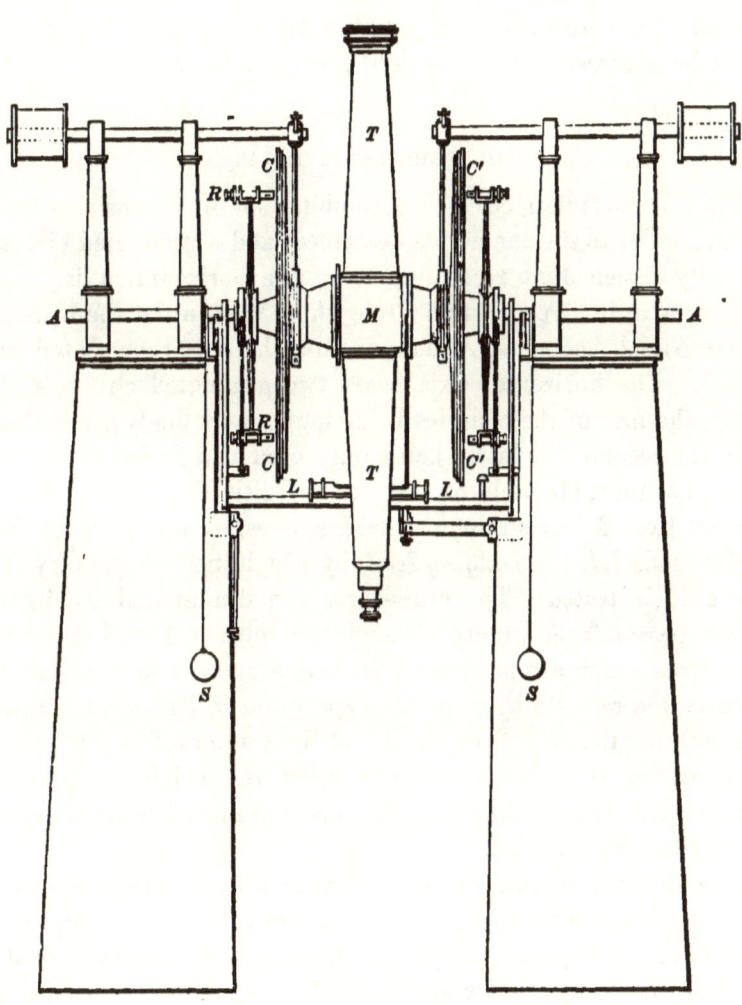

Fig 7

instrument in observatories is to the determination of the right ascensions of celestial bodies. Knowing the constant instrumental errors just now mentioned, and the error and the rate of the clock, we can easily obtain the true sidereal time at the instant of the transit of a celestial body, which time is at once, as we have already seen, the right ascension of that body.

THE MERIDIAN CIRCLE.

27. The meridian circle is a combination of a transit instrument, similar to the one above described, and a graduated circle, securely fastened at right angles to the horizontal axis, and turning with it. A meridian circle which is set up at the United States Naval Academy, Annapolis, Maryland, is represented in Fig. 7. The horizontal axis bears two graduated circles, CC, $C'C'$, the first of these circles being much more finely graduated than the second, the latter being only used as a *finder*, to set the telescope approximately at any desired altitude. R and R represent two of four stationary microscopes, by which the circle CC is read; LL is a *hanging level*, by which the horizontality of the axis is tested. The cross-wires are illuminated by light which passes from a lamp through the tubes AA, and through the pivots which are perforated for this purpose, and is reflected towards the reticule by a metallic speculum which is set within the hollow cube M. The quantity of light admitted is regulated by revolving discs with eccentric apertures, which are placed between the Ys and the tubes AA, and are moved by cords carrying small weights, SS.

The object in having so many reading microscopes is to diminish the errors arising from the imperfect graduation of the vertical circle. When any angle is to be measured, the readings of all four microscopes are first taken. The telescope, carrying the circle with it, is then moved through the angle whose value is required, and the new readings of the microscopes, which have remained stationary, are taken. We thus obtain four values, one from each microscope, of the angle measured. Theoretically, these values should be identical, and if they are not, their mean is taken as the true measure of the angle observed.

THE MICROSCOPE.

28. *The Reading Microscope.*—The reading microscope is represented in Fig. 8. The observer, placing his eye at *A*, sees the image of the divisions of the graduated circle *MN*, formed at *D*, the common focus of the glasses *A* and *C*. He will also see a scale of nótches, *nn*, and two intersecting spider threads, as shown in Fig. 9. These threads are attached to a sliding frame, *aa*, which is moved by means of a fine screw, *cc*, the head of which, *EF*, is graduated. The scale of notches is immovable, and is so constructed that the distance between the centres of any two consecutive notches is equal to that between

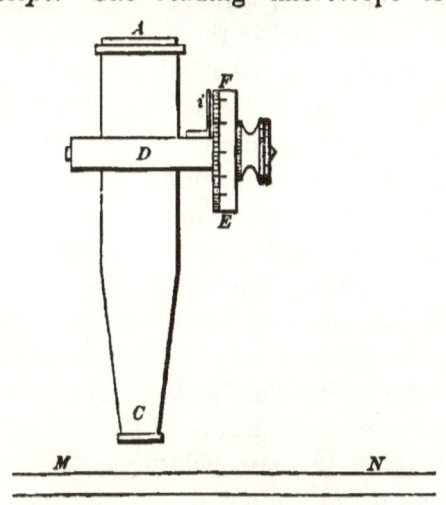

Fig. 8.

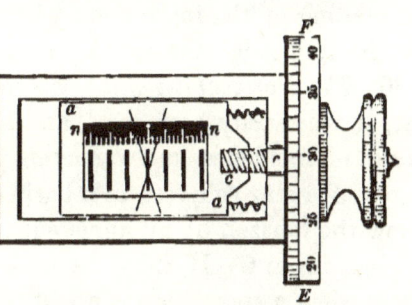

Fig. 9.

the threads of the screw, thus making the number of teeth passed over by the spider threads equal to the number of complete revolutions made by the screw. The central notch is taken as the point of reference, and is distinguished by a hole opposite to it. There is a fixed index at *i*, to which the divisions on the head of the screw are referred. When any division of the limb does not coincide with the central notch, the spider threads are moved from the central notch to the division, and the number of revolutions and fractional parts of a revolution which the screw makes is noted. If now we suppose the value of each division of the graduated circle, *MN*, to be 10', and that ten revolutions of the screw suffice to carry the spider threads across one of these divisions, then will one revolution of the screw correspond to an arc of 1'; and if we further suppose that the head of the screw is divided into 60 equal parts, then each division on the head will correspond to an arc of 1". In such a case, the complete reading of the limb is obtained to the nearest second. By increasing the power of the microscope, the fineness of the screw, and the number of the graduations on the screw-head, the reading of the limb may be obtained with far greater precision.

29. *Fixed Points.*—The meridian circle, being also a transit instrument, may be used as such; but the object for which it is specially used is the measurement of arcs of the meridian. In order to facilitate such measurement, certain *fixed points* of reference are determined upon the vertical circle. The most important of these points are the *horizontal point*, by which is meant the reading of the instrument when the axis of the telescope lies in the plane of the horizon; the *polar point*, which is the reading of the instrument when the telescope is directed to the elevated pole; the *zenith point*, and the *nadir point*.

30. *The Horizontal Point.*—As the surface of a fluid, when at rest, is necessarily horizontal, and as, by the laws of Optics, the angles of incidence and reflection are equal to each other, the image of a star reflected in a basin of mercury will be depressed below the horizon by an angle equal to the altitude of the star at that instant. If, then, we take the reading of the vertical circle when a star which is about to cross the meridian is on the first vertical thread of the reticule, and then, depressing the telescope, take the reading of the circle when the reflected image of the star crosses the last vertical thread, and, by means of small corrections, reduce these readings to what they would have been, had both star and image been on the meridian when the observations were made, the mean of these two reduced readings will be the *horizontal point*. The horizontal point having been thus determined, the *zenith point* and the *nadir point*, being situated at intervals of 90° from it, are at once obtained. Knowing the horizontal and the zenith point, we are able to measure the meridian altitude or the meridian zenith distance of any celestial body which comes above our horizon. And further, as the latitude is equal to the altitude of the elevated

pole, if the latitude of the place is accurately known, we can at once obtain the *polar point* by applying the latitude to the horizontal point.

31. *Nadir Point.*—The nadir point may be independently obtained in the following manner. Let the telescope, represented in Fig. 10 by AB, be directed vertically downwards towards a basin of mercury, CD. The observer, placing his eye at A, will see the cross-wires of the telescope, and will see also the *image* of these wires reflected into the telescope from the mercury. By slowly moving the telescope, the cross-wires and their reflected image may be brought into exact coincidence, and the reflected image will then disappear. The line of sight of the telescope is now vertical, and the reading of the vertical circle will be the nadir point, from which the other points can readily be found.

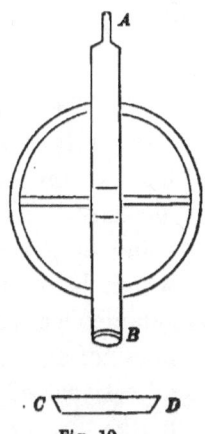

Fig. 10.

There is a variety of methods by which each of these points can be obtained, without reference to any other; and by comparing the results which these different independent methods give, the errors to which each result is liable may be very considerably diminished.

32. *Use of the Meridian Circle.*—The meridian circle may be used in connection with the sidereal clock, to find the right ascension and declination of any celestial body. The telescope is directed towards the body as it crosses the meridian, and the time of transit as shown by the clock, and the reading of the vertical circle, are both taken. We have already seen how the right ascension of the body is obtained from the clock time of transit. The difference between the reading of the circle, which we suppose to have been taken, and the polar point, is the polar distance of the body, the complement of which is the declination. Or we may obtain the declination still more directly by previously establishing the *equinoctial point* of the instrument, the reading, that is to say, of the vertical circle when the telescope lies in the plane of the equinoctial.*

---

* As the direction in which a star *appears* to lie is not, owing to refrac-

On the other hand, if we make the same observations upon a star whose right ascension and declination are known, we can determine the latitude and the sidereal time of the place of observation.

THE MURAL CIRCLE.

33. The *mural circle* is, in construction, adjustment, and use, essentially a meridian circle. The only important difference between the two instruments is in the manner in which they are mounted. The horizontal axis of the mural circle, instead of being supported at both extremities, is supported only at one, which is let into a stone pier or wall. Owing to the lack of symmetrical support, and also to the fact that the instrument does not admit of *reversal* (which is an important element in the adjustment of the meridian circle, and consists in lifting it out of the Ys, and turning the horizontal axis end for end), the mural circle can be regarded only as an inferior type of the meridian circle.

THE ALTITUDE AND AZIMUTH INSTRUMENT.

34. The general principles on which the *altitude and azimuth instrument* is constructed are seen in Fig. 11. Through the centre of a graduated circle, $C'C'$, and perpendicular to its plane, is passed an axis, AA. At right angles to this axis is a second axis, one extremity of which is represented by B. This second axis carries the telescope, TT, and also a second graduated circle, CC, whose plane is perpendicular to that of the circle $C'C'$. The telescope admits of being moved in the plane of each circle, and

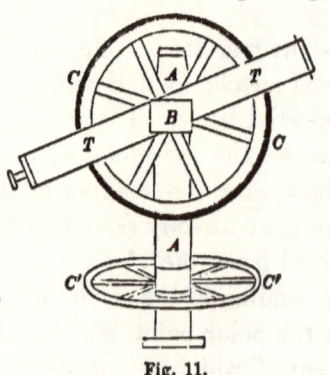

Fig. 11.

---

tion and other causes, which will be explained in Chap. III., the direction in which it really lies, certain small corrections must be applied to the reading of the circle to obtain the reading which really corresponds to the direction of the star.

microscopes or verniers are attached to the instrument, by means of which arcs on either circle can be read.

If this instrument is so placed that the principal axis, AA, lies in a vertical direction, we shall have an altitude and azimuth instrument, sometimes called an *altazimuth*. The circle $C'C'$ will then lie in the plane of the horizon, and the axis AA, indefinitely prolonged, will meet the surface of the celestial sphere in the zenith and the nadir. The circle CC, as it is moved with the telescope about the axis AA, will continually lie in a vertical plane.

35. *Fixed Points.*—Altitudes may be measured on the vertical circle, when we know the horizontal or the zenith point, the determination of which has already been described in Arts. 30 and 31. In order to measure the azimuth of any celestial body, we must, in like manner, establish some fixed point of reference on the horizontal circle, as, for instance, the north or the south point, by which is meant the reading of the horizontal circle when the telescope lies in the plane of the meridian.

36. *Method of Equal Altitudes.*—One of the most accurate methods of obtaining the north or the south point of the horizontal circle is called the *method of equal altitudes*. Let Fig. 12 represent the projection of the celestial sphere on the plane of the celestial horizon, $NESW$. Z is the projection of the zenith, P of the pole, and the arc AA' the projection of a portion of the diurnal circle of a fixed star, which is supposed to have the same altitude when it reaches A', west of the meridian, which it had at A, east of the meridian.

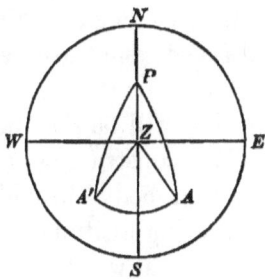

Fig. 12.

Now, in the two triangles PAZ, $PA'Z$, we have PZ common, PA' equal to PA (since the polar distance of a fixed star remains constant), and ZA equal to ZA', by hypothesis; the two triangles are therefore equal in all their parts, and hence the angles PZA and PZA' are equal. But these two angles are the azimuths of the star at the two positions A and A'. We may say, then, in general, that equal altitudes of a fixed star correspond to equal angular distances from the meridian.

Now, let the telescope be directed to some fixed star east of the meridian, and let the reading of the horizontal circle be taken. When the star is at the same altitude, west of the meridian, let the reading of the horizontal circle again be taken; the mean of these two readings is the reading of the horizontal circle when the axis of the telescope lies in the plane of the meridian.

37. *Use of the Altitude and Azimuth Instrument.*—This instrument is chiefly used for the determination of the amount of refraction corresponding to different altitudes. Refraction, as will be seen in the next chapter, displaces every celestial body in a vertical direction, making its apparent zenith distance less than its true zenith distance. At the instant of taking the altitude of a celestial body, the local sidereal time is noted, from which, knowing the right ascension of the body, we can obtain its hour angle from the theorem in Art. 9. We shall then have in the astronomical triangle, PZA (Fig. 3), the hour angle ZPA, the side PA, or the polar distance of the body, and the side PZ, the co-latitude of the place of observation. We can, therefore, compute the side ZA, which is the *true* zenith distance of the body observed; and the difference between this and the *observed* zenith distance, (corrected for parallax, [Art. 55,] and instrumental errors,) will be the amount of refraction for that zenith distance.

The construction of the instrument enables us to follow a celestial body through its whole course from rising to setting, measuring altitudes and noting the corresponding times to any extent that we choose; and the amount of refraction corresponding to each altitude can afterwards be computed at our leisure.

THE EQUATORIAL.

38. The *equatorial* is similar in general construction to the altitude and azimuth instrument. It is, however, differently placed, the plane of the principal graduated circle, $C'C'$, coinciding, not with the plane of the horizon, but with that of the celestial equator, from which peculiarity of position comes the

name of *equatorial*. The circle $C'C'$, when thus placed, is called the *hour circle* of the instrument, and the axis A, at right angles to it, is called the *hour* or *polar axis*. It is evident that the axis is directed towards the poles of the heavens. The circle CC, the plane of which is perpendicular to the plane of the circle $C'C'$, will lie continually in the plane of a circle of declination, as the instrument is turned about the polar axis, and is hence called the *declination circle*. The axis on which this latter circle is mounted is called the *declination axis*.

39. *Use of the Equatorial.*—The equatorial is employed principally in that class of observations which require a celestial body to remain in the field of view during a considerable length of time. The manner in which this requirement is met is explained in Fig. 13. Let AA' be the polar axis of an equatorial, directed towards the pole of the heavens, P. Let $ss's''$ be the diurnal circle in which a star appears to move about the pole. Suppose the telescope, TT, to be turned in the direction of the star when at s, and

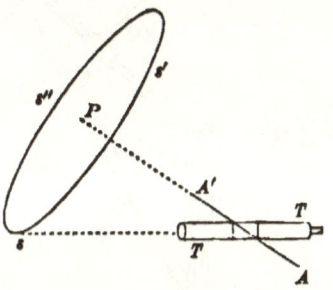

Fig. 13.

to be moved until the intersection of the cross-wires and the star coincide, and then clamped. Now, if the instrument is made to revolve about the axis AA', with an angular velocity equal to that of the star about the pole, it is plain that, since the angle which the axis of the telescope makes with the polar axis remains unchanged, and is continually equal to the angular distance of the star from the pole, the coincidence of the cross-wires and the star will remain complete.

A clock-work arrangement, called a *driving-clock*, is now usually connected with large equatorials, by which the instrument may be moved uniformly about its polar axis, at the required rate, so that the observer has ample time to measure the angular diameter of a celestial body, to measure the angular distance between two stars which are near each other, and to make other micrometric observations of a similar character.

THE SEXTANT.

40. The *sextant* is an instrument by which the angular distance between two visible objects may be measured. It is used chiefly by navigators; but its portability gives it great value wherever celestial observations are required. The angles for the measurement of which it is used are the altitudes of celestial bodies, and the angular distances between celestial bodies or terrestrial objects.

Fig. 14 is a representation of the sextant. Its form is that

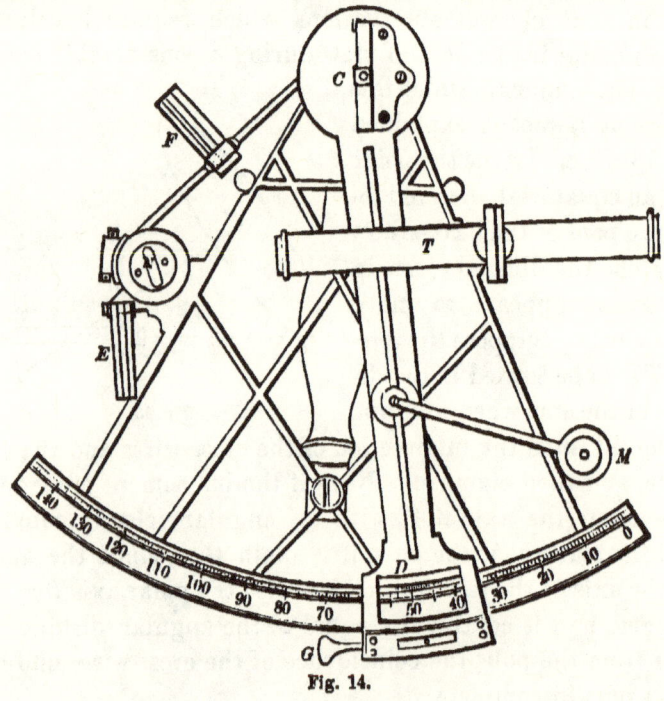

Fig. 14.

of a sector of a circle, the arc of which comprises 60°. A movable arm, *CD*, called the *index-bar*, revolves about the centre of the sector. This bar carries at one extremity a vernier, *D*. At the other extremity of the index-bar, and revolving with it, is placed a silvered mirror, *C*, the surface of which must be perpendicular to the plane of the instrument. This glass is called the *index-glass*. Another glass, *N*, called the *horizon-glass*, is attached to the frame of the instrument, and only its lower half is silvered.

This glass is immovable, and its surface must be perpendicular to the plane of the instrument. *T'* is a telescope, directed towards the horizon-glass, with its line of sight parallel to the plane of the instrument. *F* and *E* are two sets of colored glasses, which may be used to protect the eye when the sun is observed. *M* is a magnifying glass, to assist the eye in reading the vernier. *G* is a tangent screw, which gives a slight motion to the index-bar, and is used in obtaining an accurate coincidence of the images.

41. *Optical Principle of Construction.*—The sextant is constructed upon a principle in Optics which may be stated thus:— *The angle between the first and the last direction of a ray which has suffered two reflections in the same plane is equal to twice the angle which the two reflecting surfaces make with each other.*

To prove this: In Fig 15, let *A* and *B* be the two reflecting surfaces, supposed to be placed with their planes perpendicular to the plane of the paper. Let *SA* be a ray of light from some body, *S*, which is reflected from *A* to *B*, and from *B* in the direction *BE*. Prolong *SA* until it meets the line *BE*. Then will the

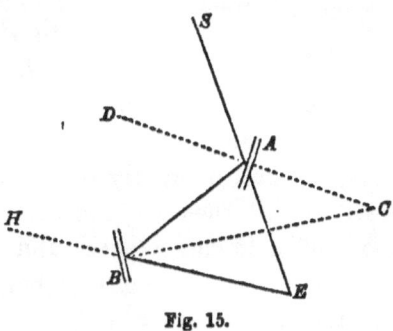

Fig. 15.

angle *SEB* be the angle between the first and the last direction of the ray *SA*. At the points *A* and *B* let the lines *AD* and *BC* be drawn perpendicular to the reflecting surfaces, and prolong *AD* until it meets *BC*. The angle *DCB* is equal to the angle which the two surfaces make with each other. We have then to prove that the angle *SEB* is double the angle *DCB*.

Now since the angle of incidence always equals the angle of reflection, *SAD* and *DAB* are equal, and so are *ABC* and *CBE*. We have, by Geometry,

$$SEB = SAB - ABE$$
$$= 2(DAB - ABC),$$
$$= 2\,DCB.$$

42. *Measurement of Angular Distances.*—Suppose, now, that

we wish to measure the angular distance between two celestial bodies, A and B (Fig. 16). The instrument is so held that its plane passes through the two bodies, and the fainter of them, which in this case we suppose to be B, is seen directly through the horizon-glass and the telescope. B is so distant that the rays $B'C$ and Bm, coming from it, may be considered to be sensibly parallel. Let ab and CI be the positions of the index-glass and index-bar when the index-glass and the horizon-glass are parallel. Then will the ray $B'C$ be reflected by the two glasses in a direction parallel to itself, and the observer, whose eye is at D, will see both the direct and the reflected image of B in coincidence. Now let the index-bar be moved to some new position, CI', so that the ray from the second body, A, shall be finally reflected in the direction of mD. The observer will then see the direct image of B and the reflected image of A in coincidence; and the angular distance between the two bodies is evidently equal to the angle between the first and the last direction of the ray AC, which angle has already been shown to be equal to *twice* the angle which the two glasses now make with each other, or to twice the angle ICI'. If, then, we know the point I on the graduated arc at which the index-bar stands when the glasses are parallel, twice the difference between the reading of that point and that of the point I' will be the angular distance of the two bodies.

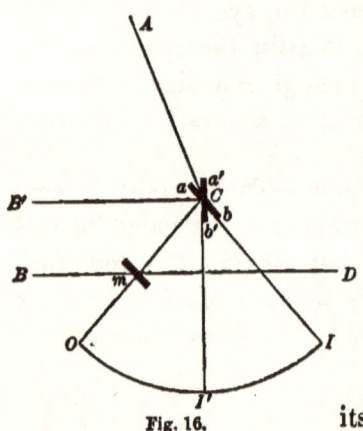

Fig. 16.

To avoid this doubling of the angle, every *half* degree of the arc is marked as a *whole* degree, when the graduation is made; so that, in practice, we have only to subtract the reading of I from that of I' to obtain the angle required.

43. *Index Correction.*—The point of reference on the arc from which all angles are to be reckoned is, as we have already seen, the reading of the sextant when the surfaces of the index-glass and the horizon-glass are parallel. This point may fall either

at the zero of the graduation, or to the left or to the right of it; and to provide for the last case, the graduation is carried a short distance to the right of the zero, this portion of the arc being called the *extra arc*. The reading of this point of parallelism is called the *index correction*, and is *positive* when it falls to the *right* of the zero, and negative when it falls to the left. Suppose, for instance, that the instrument reads 2' on the extra arc when the glasses are parallel: all angles ought then to be reckoned from the point 2', instead of from the zero point; in other words, 2' is a *constant correction to be added to every reading*.

There are several methods of finding the index correction. One method, which can readily be shown from Fig. 16 to be a legitimate one, is to move the index-bar until the direct and the reflected image of the same star are in coincidence, and then take the reading, giving it its proper sign according to the rule above stated. Another method, generally more convenient, in which the sun is used, may be found in Bowditch's Navigator, and in most treatises on Astronomy: where also may be found the methods of testing the adjustments of the sextant.

44. *The Artificial Horizon.*—In order to obtain the altitude of a celestial body at sea, the sextant is held in a vertical position, and the index-bar is moved until the reflected image of the body is brought into contact with the visible horizon seen through the telescope of the sextant. The sextant reading is then corrected for the index correction; and corrections must also be applied for parallax, refraction, and the dip of the horizon, as will be explained in the next Chapter. If the body observed is the sun or the moon, either its upper or its lower limb is brought into contact with the horizon, and the value of its angular semi-diameter (given in the Nautical Almanac) is subtracted or added.

On shore, use is made of the *artificial horizon*, already alluded to in Art. 30. This commonly consists of a shallow, rectangular basin of mercury, the surface of which is protected from the wind by a sloping roof of glass. The observer so places himself that he can see the image of the body whose altitude he wishes to measure reflected in the mercury. He then moves the index-bar of the sextant until the image of the body reflected by the sextant is in coincidence with that reflected by the mercury. The sextant reading is then corrected for the *whole* of the index correction. Half of the result will be, as shown in Art. 30, the apparent altitude of the body, to which must be applied the corrections for parallax and refraction to obtain the true altitude. When the sun or the moon is observed, the upper or the lower limb of the image reflected by the sextant is brought into contact with the opposite limb of the image reflected by the mercury, and the correction for semi-diameter also is applied.

45. *The Vernier.*—The *vernier* is an instrument by which, as by the reading microscope previously explained, fractions of a division of a limb may be read.

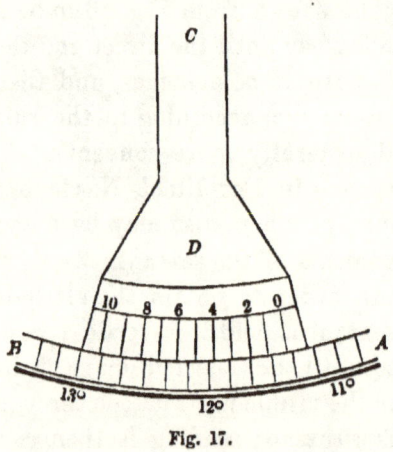

Fig. 17.

In Fig. 17, let AB be an arc of a stationary graduated circle, and let CD be a movable arm, carrying another graduated arc at its extremity. The value of each division of the limb AB is one-sixth of a degree, or 10'. The arc on the arm CD is divided into *ten* equal parts, and the length of the arc between the points 0 and 10 is equal to the length of *nine* divisions of the arc AB. This arc, which the limb CD carries, is called a *vernier*. Since the ten divisions of the vernier equal in length nine divisions of the limb, it follows that each division of the vernier comprises 9' of arc; in other words, any division of the vernier is less by 1' of arc than any division of the limb.

The reading of any instrument which carries a vernier is always determined by the position of the zero point of the vernier. If, now, the zero point of the vernier exactly coincides with a division of the limb, the point 1 of the vernier will fall 1' behind the next division of the limb, the point 2 will fall 2' behind the next division but one, and so on; and if, such being the case, the vernier is moved forward through an arc of 1', the point 1 will come into coincidence with a division of the limb; if it is moved forward through an arc of 2', the point 2 will come into coincidence with a division on the limb; and, in general, the number of minutes of arc by which the zero point of the vernier falls beyond the division of the limb which immediately precedes it will be equal to the number of that point of the vernier which is in coincidence with a division of the limb. If, then, the zero point falls between any two divisions of the limb, as 11° 20′ and 11° 30′, for example, and the point 2 of the vernier is found to be in coincidence with any division of the limb, we know that the zero point is 2′ beyond the division 11° 20′, and that the complete reading for that position of the vernier is 11° 22′.

46. *General Rules of Construction.*—In the construction of all verniers similar to the one above described, the same rules of construction must be followed: the length of the arc of the vernier must be exactly equal to the length of a certain number (no matter what) of the divisions of the limb, and the arc must be divided into equal parts, the number of which shall be greater by one than the number of these divisions of the limb. Following these rules, and putting
$D =$ the value of a division of the limb,
$d =$ "   "   "   "   "   vernier,
$n =$ the number of equal parts into which the vernier is divided,
we have $\qquad D - d = \dfrac{D}{n} \qquad$ as a general formula.

The difference $D - d$ is called the *least count* of the vernier.

If, in Fig. 17, we take the length of the vernier equal to 59 divisions of the limb, and divide it into 60 equal parts, we shall have
$$D - d = \frac{10'}{60} = 10'':$$
which is the least count on most of the modern sextants.

Verniers are sometimes constructed in which the number of equal parts on the vernier is *less* by one than the number of the divisions of the limb taken. In this case we have $d - D = \dfrac{D}{n}$: and the only difference between this class of verniers and the class above described is that the graduations of the limb and the vernier proceed in this class in opposite directions.

OTHER ASTRONOMICAL INSTRUMENTS.

47. The *zenith telescope*, the *theodolite*, and the *universal instrument* are, in general principle, only modified forms of the portable altitude and azimuth instrument.

The *octant* (sometimes improperly called the *quadrant*) is identical in construction with the sextant, excepting only that its arc contains 45°.

The *prismatic sextant* carries a reflecting prism in place of the ordinary horizon-glass, and the graduated arc comprises a semi-circumference.

The *reflecting circle* is still another modification of the sextant, in which the graduated arc is an entire circumference, and the index-bar is a diameter of the circle, revolving about the centre, and carrying a vernier at each extremity. Sometimes the circle has three verniers, at intervals of 120° of the graduated arc.

The *spectroscope* is an instrument which is used, as its name indicates, in the examination of the spectra both of terrestrial substances and of the heavenly bodies. Its use as an instrument of astronomical research is comparatively recent, but it has already led to many interesting and remarkable discoveries concerning the constitution of the heavenly bodies. It consists essentially of three parts: a tube, a prism (or a set of prisms), and a telescope. Rays of light from either a celestial body or an artificial flame are made to enter the tube through an extremely narrow slit at its extremity. These rays pass through the tube, and fall upon the prism. If necessary, lenses may be placed within the tube, so that the rays, as they issue from it, shall fall upon the prism in parallel lines. These rays are dispersed by the prism, and a spectrum is formed. This spectrum is then examined by means of the telescope. There is also an arrangement by which rays of light from two substances or bodies can be introduced through the slit without interfering with each other, so that their spectra can be formed simultaneously, one above the other, and the points of resemblance or difference between them can be accurately noted.

It is well known that the solar spectrum contains a large

number of dark and narrow parallel lines, which are called Fraunhofer's lines. The spectra of the stars and of artificial lights also contain similar series of lines, differing from each other, each series, however, being constant for the same body or light. The spectra of chemical substances also present certain peculiarities, so that each spectrum indicates with certainty the substance which produces it. Hence, by a comparison of the spectra of the heavenly bodies with those of known chemical substances, the existence of many of those substances in the heavenly bodies has been definitely established. Nor is this all; the inspection of any spectrum suffices to tell us whether the light which forms it comes from a solid or a gaseous body, and whether, if the light comes from a solid body, it passes through a gaseous body before it reaches us.

The results of these investigations will be noticed when we come to the description of the heavenly bodies; and the method of investigation will be further illustrated in the Article on the constitution of the sun. (Art. 102.)

ERRORS.

48. However carefully an instrument may be constructed, however accurately adjusted, and however expert the observer may be, every observation must still be regarded as subject to errors. These errors may be divided into two classes, *regular* and *irregular errors*. By *regular errors* we mean errors which remain the same under the same combination of circumstances, and which, therefore, follow some determinate law, which may be made the subject of investigation. Among the most important of this class of errors are *instrumental errors:* errors, that is to say, due to some defect in the construction or adjustment of an instrument. If, for instance, what we call the vertical circle of the meridian circle is not rigorously a circle, or is imperfectly graduated; or if the horizontal axis is not exactly horizontal, or does not lie precisely east and west; any one of these imperfections will affect the accuracy of the observation. The observer, however, knowing what the construction and adjustment of the instrument ought to be, can calculate what effect any given im-

perfection will produce upon his observation, and can thus determine what the observation would have been had the imperfection not existed. *Regular* errors, then, may be neutralized by determining and applying the proper *corrections*.

Irregular errors, on the contrary, are errors which are not subject to any known law. Such, for example, are errors produced in the amount of refraction by anomalous conditions of the atmosphere; errors produced by the anomalous contraction or expansion of certain parts of the instrument, or by an unsteadiness of the telescope produced by the wind; and, more particularly, errors arising from some imperfection in the eye or the touch of the observer. Errors such as these, being governed by no known law, can never be made the subject of theoretic investigation; but being by their very nature accidental, the effects which they produce will sometimes lie in one direction and sometimes in another; and hence the observer, by repeating his observations, by changing the circumstances under which he makes them, by avoiding unfavorable conditions, and finally by taking the *mean*, or the *most probable* value of the results which his different observations give him, can very much diminish the errors to which any single observation would be exposed.

NOTE.—For complete descriptions of the various astronomical instruments, the student is referred to Chauvenet's *Spherical and Practical Astronomy;* Loomis's *Practical Astronomy;* and Pearson's *Practical Astronomy* (published in England).

CHAPTER III.

REFRACTION. PARALLAX. DIP OF THE HORIZON.

REFRACTION.

49. When a ray of light passes obliquely from one medium to another of different density, it is bent, or *refracted*, from its course. If a line is drawn perpendicular to the surface of the second medium at the point where the ray meets it, the ray is bent towards this perpendicular if the second medium is the denser of the two, and from it if the first medium is the denser.

In Fig. 18, let AA, BB, represent two media of different density, the density of BB being the greater. Let CD be a ray of light meeting the surface of BB at D. At D erect the line ND perpendicular to the surface of BB, and prolong it in the direction DM. The ray CD is

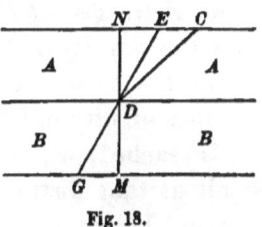

Fig. 18.

called the *incident ray*, and the angle NDC the *angle of incidence*. When the ray enters the medium BB, it will still lie in the same plane with CD and ND, but will be bent towards the line DM, making with it some angle GDM, less than the angle NDC. To an observer whose eye is at G, the ray will appear to have come in the direction EG, which is therefore called the *apparent direction* of the ray. DG is called the *refracted ray*, and the angle GDM the *angle of refraction*. The angle EDC, the difference between the directions of the incident and the refracted ray, is called the *refraction*.

. It is shown in Optics that, whatever the angle of incidence may be, there always exists a constant ratio between the sine of the angle of incidence and that of the angle of refraction, as long as the same two media are used and their densities are unchanged. We have, then, in the figure,—

$$\frac{\sin NDC}{\sin GDM} = k:$$

k being a constant for the two media AA and BB.

If the second medium, instead of being of uniform density, is composed of parallel strata, each one of which is of greater density than the one immediately preceding, as is represented in Fig. 19, the path of the ray through these several strata will be a broken line, $Dabc$; and if the thickness of each of these successive strata is supposed to be indefinitely small, this broken line will become a curve.

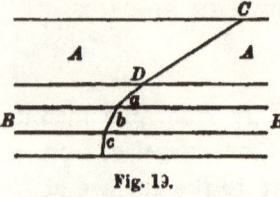

Fig. 19.

In the figures above used, the media are represented as separated by plane surfaces; but the same phenomena are noticed, and the same laws hold good, if the media are separated by curved surfaces.

50. *Astronomical Refraction.*—It is determined by experiment that the density of the air gradually diminishes as we ascend above the surface of the earth, and it is estimated that at a distance of fifty miles above the surface the upper limit of the air is reached; or, at all events, that the density of the air is so small at that distance that it exerts no appreciable refracting power. We may, therefore, consider the air to be made up of a series of strata concentric with the earth's surface, the thickness of each stratum being indefinitely small, and the density of each stratum being greater than that of the stratum next above it. Now, in Fig. 20, let the arc BD represent a portion of the earth's surface, and the arc MN a portion of the upper limit of the atmosphere. Let S be a celestial body, and SA a ray of light from it, which enters the atmosphere at A. Let the normal (or radius) AC be

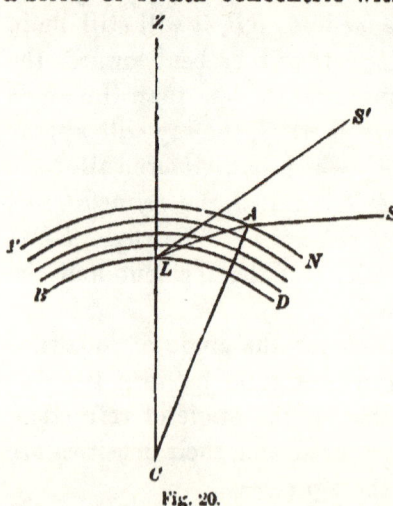

Fig. 20.

drawn. As the ray of light passes down through the atmosphere, it is continually passing from a rarer to a denser medium, so that its path is continually changed, and becomes a curve AL, concave towards the earth, and reaching the earth at some point L. Since the direction of a curve at any point is the direction of the tangent to the curve at that point, the apparent direction of the ray of light at L will be represented by the tangent LS', and in that direction will the body S appear to lie, to an observer at L. If the radius CL be indefinitely prolonged, the point Z, where it reaches the celestial sphere, will be the zenith of the observer at L, the angle ZLS' will be the *apparent* zenith distance of the body S, and the angle which the line drawn from the body to the point L makes with the line LZ will be the *true* zenith distance of S. The effect, then, of refraction is to decrease the apparent zenith distances, or increase the apparent altitudes of the celestial bodies. Since the incident ray SA, the curve AL, and the tangent LS' all lie in the same vertical plane, the azimuth of the celestial bodies is not affected.

51. *General Laws of Refraction.*—By an investigation of the formulæ of refraction, and by astronomical observations already described (Art. 37), the amount of refraction at different altitudes has been obtained, and is given in what are called "tables of refraction." The following general laws of refraction will serve to give the student some idea of its amount, and of the conditions under which it varies:—

(1.) In the zenith there is no refraction.

(2.) The refraction is at its maximum in the horizon, being there equal to about 33'. At an altitude of 45° it amounts to 57".

(3.) For zenith distances which are not very large, the refraction is nearly proportional to the tangent of the zenith distance. When the zenith distance is large, however, the expression of the law is much more complicated. No table of refraction can be trusted for an altitude of less than 5°.

(4.) The amount of refraction depends upon the density of the air, and is nearly proportional to it. The tables give the refraction for a mean state of the atmosphere, taken with the barometer at 30 inches and the thermometer at 50°. If the temperature remains constant, and the barometer stands above its mean height, or if the height of the barometer is constant, and the thermometer stands below its mean height, the density of the atmosphere is increased, and the refraction is greater than its mean amount. Supplementary tables are therefore given, from which, with the observed heights of both barometer and thermometer as arguments, we may take the necessary corrections to be applied to the mean refraction.

(5.) Since the effect of refraction is to increase the apparent altitudes of the celestial bodies, the amount of refraction for any apparent altitude is to be subtracted from that apparent altitude, or added to the corresponding zenith distance.

52. *Effects of Refraction.*—The apparent angular diameter of the sun and of the moon being about 32', and the refraction in the horizon being 33', it follows that when the lower limb of either body appears to be resting on the horizon, the body is in reality below it. One effect, then, of refraction is to lengthen the time during which these bodies are visible. Still another effect is to distort the discs of the sun and the moon when near the horizon: for since the refraction varies rapidly near the horizon, the lower extremity of the vertical diameter of the body will be more raised than the upper extremity, thus apparently shortening this diameter, and giving the body an elliptical shape. When the body comes still nearer the horizon, its disc is distorted into what is neither a circle nor an ellipse, but a species of oval, in which the curvature of the lower limb is less than that of the upper one. The apparent enlargement of these bodies when near the horizon is merely an optical delusion, which vanishes when their diameters are measured with an instrument.

PARALLAX.

53. The *parallax* of any object is, in the general sense of the word, the difference of the directions of the straight lines drawn to that object from two different points: or it is the angle at the object subtended by the straight line connecting these two points. In Astronomy, we consider two kinds of parallax: *geo-*

centric parallax, by which is meant the difference of the directions of the straight lines drawn to the centre of any celestial body from the earth's centre and any point on its surface, and *heliocentric* parallax, or the difference of the directions of the lines drawn to the centre of the body from the centre of the earth and the centre of the sun. The former is the angle at the body subtended by that radius of the earth which passes through the place of observation: the latter the angle at the body subtended by the straight line joining the centre of the earth and that of the sun.

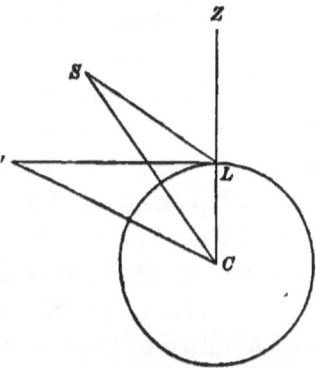

Fig. 21.

54. *Geocentric Parallax.*—In Fig. 21, let C be the centre of the earth, and L some point on its surface, of which Z is the zenith. Let S be some celestial body. The geocentric parallax of the body is the angle CSL. Let S' be the same body in the horizon. The angle $LS'C$ is the parallax of the body for that position, and is called its *horizontal parallax*. If we denote this horizontal parallax by P, the earth's radius by R, and the distance of the body from the earth's centre by d, we have, by Trigonometry,

$$\sin P = \frac{R}{d}.$$

To find the parallax for any other position, as at S, we represent the angle LSC by p, and the apparent zenith distance of the body, or the angle ZLS, by z, the sine of which is equal to the sine of its supplement SLC. We have from the triangle LSC, since the sides of a plane triangle are proportional to the sines of their opposite angles,

$$\frac{\sin p}{\sin z} = \frac{R}{d}.$$

Combining this equation with the preceding, we have,

$$\sin p = \sin P \sin z.$$

Since P and p are small angles, we may consider them proportional to their sines, and thus have, finally,

$$p = P \sin z.$$

The parallax, then, is proportional to the sine of the zenith distance, and may be found for any altitude when the horizontal parallax is known. It evidently decreases as the altitude increases, and in the zenith becomes zero.

55. *Application of Parallax.*—In order that observations made at different points of the earth's surface may be compared, they must be reduced to some common point. Geocentric parallax is applied to reduce any altitude observed at any place to what it would have been had it been observed at the earth's centre. We see from Fig. 21 that parallax acts in a vertical plane, and that the zenith distance of the body as observed from the earth's centre, or the angle ZCS, is less than the observed zenith distance ZLS, by the parallax CSL. Parallax, then, is always subtractive from the observed zenith distance, and additive to the observed altitude.

The parallax above described is, strictly speaking, the parallax *in altitude*. There is also, in general, a similar parallax *in right ascension*, and *in declination*, formulæ for deriving which from the parallax in altitude are given in other works.

56. *Heliocentric Parallax.*—It may sometimes happen that we wish to reduce an observation from what it was at the centre of the earth to what it would have been if it had been made at the centre of the sun. Fig. 21, and the formulæ obtained from it, will apply equally well to this case, by making the necessary changes in the description of the figure and in the names of the angles.

Let S be still a celestial body, but let C be the centre of the sun, and L that of the earth. The angle p will then represent the heliocentric parallax, and the angle SLC the angular distance of the body from the sun, as measured from the earth's centre, or, as it is called, the body's *elongation*. The angle Γ will be the greatest value of the heliocentric parallax, taken when the body's elongation from the sun is 90°, and is called the *annual* parallax. We shall then find, from the formulæ of Art. 54, that the annual parallax has for its sine the ratio of the distance of the earth from the sun to that of the body from the sun, and that the parallax for any other position is the product of the annual parallax by the sine of the body's elongation.

DIP OF THE HORIZON.

57. The *dip of the horizon* is the angular depression of the visible horizon below the celestial horizon. In Fig. 22, let HG be a portion of the earth's surface, and C the earth's centre. Let a radius of the earth, CA, be prolonged to some point D, beyond the surface, and let an observer be supposed to be at the point D. At the point D let the line BD be drawn perpendicular to the line CD, and also the line DH, tangent to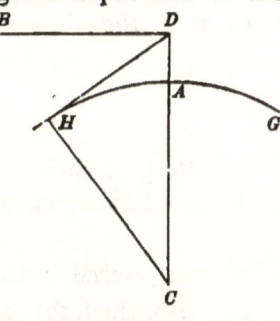

Fig. 22.

the earth's surface at some point H. If these two lines be revolved about the line CD, DB will generate the plane of the celestial horizon (since we have seen that all planes passed perpendicular to the radius will, when indefinitely extended, mark out the same great circle on the celestial sphere), and DH will generate the surface of a cone, which will touch the earth in a small circle. If we disregard for the present the effect of the earth's atmosphere, this small circle will be the visible horizon of the observer at D, and the angle BDH will be the dip. DA is the linear height of the observer at D.

Now let a radius, CH, be drawn to the point of tangency H. The angles BDH and HCD, having their sides mutually perpendicular, are equal. Represent the angle HCD by D, the earth's radius by R, and the observer's height, AD, by h. In the triangle DHC, right-angled at H, we have,

$$\frac{R}{R+h} = \cos D:$$

since D is small, it is better to use the formula,

$$\cos D = 1 - 2\sin^2 \tfrac{1}{2} D:$$

hence we have,

$$\frac{R}{R+h} = 1 - 2\sin^2 \tfrac{1}{2} D:$$

$$\therefore \sin \tfrac{1}{2} D = \sqrt{\frac{h}{2(R+h)}}.$$

As the angle D is small, we may take
$$\sin \tfrac{1}{2} D = \tfrac{1}{2} D \sin 1'';$$
and as h is very small in comparison with R, we may also assume $(R+h)$ to be sensibly equal to R. Making these changes in the above equation, and finding the value of D, we have
$$D = \frac{2}{\sin 1''} \sqrt{\frac{h}{2R}}$$
Substituting in this expression the value of the earth's radius in feet, we have, finally,
$$D = 63''.8 \sqrt{h,}$$
h being expressed in feet.

The dip, then, for a height of one foot is $63''.8$; and for other heights it is proportional to the square root of the number of feet in the height.

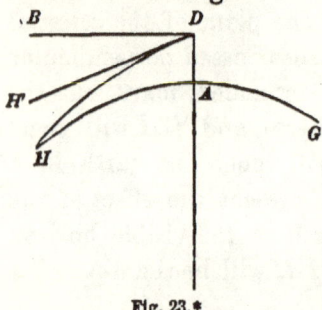

Fig. 23.*

58. *Effect of Atmospheric Refraction.*—If the effect of atmospheric refraction is taken into consideration, the line HD must be a curved line, as is represented in Fig. 23. The point H will then appear to lie in the direction DH', to the observer at D, and the dip will be the angle BDH'. The effect, then, of refraction is to *decrease* the dip, the amount by which it is decreased being about $\tfrac{1}{13}$th of the whole.

59. *Application of Dip.*—Tables have been computed in which may be found the proper dip for different heights above the surface of the earth. Dip constitutes one of the corrections which are to be applied at sea to the observed altitude of a celestial body to obtain its true altitude: its altitude, that is, above the celestial horizon. Since the visible horizon lies below the celestial horizon, this correction is evidently *subtractive*.

---

* The curved line HD is tangent to the earth's surface at H.

CHAPTER IV.

THE EARTH. ITS SIZE, FORM, AND ROTATION.

60. HAVING seen what are the construction and the adjustments of the principal astronomical instruments, to what uses each is adapted, and what corrections are to be applied to the observations which are taken, we are now ready to proceed to the solution of some of the many questions of interest which the study of Astronomy opens to us. And first of all, let us see how astronomical observations will help us to a knowledge of the size and the form of the earth. The question is one of the first importance; for upon the determination of the size of the earth depends in a great measure, as we shall see further on, the determination of the magnitudes and the distances of the other heavenly bodies. Having seen the facts which seem to point to the conclusion that the earth is spherical in form, we will start with the assumption that this conclusion is correct, and proceed to determine the magnitude of the earth, regarded as a sphere.

Now, we know from Geometry what is the ratio between the radius of a sphere and the circumference of any great circle of that sphere; and, therefore, if we can obtain the length of the circumference of any great circle of the earth, of a meridian, for instance, we can at once determine the radius of the earth. And more than this: if we can measure the length of any known arc of this meridian, of one degree, for instance, we can compute the length of the entire circumference. The determination of the earth's radius, then, depends only on our ability to satisfy these two conditions:

(1.) We must be able to measure the *linear* distance on the earth's surface between two points on the same meridian.

(2.) We must be able to measure the *angular* distance between these same two points.

61. *First Condition.*—A reference to Fig. 24 will explain how

the first of these two conditions may be satisfied. Let A and G represent the two points on the same meridian the distance between which we wish to measure. We have already seen (Art. 36) how an altitude and azimuth instrument may be so adjusted that the sight-line of the telescope will lie in the plane of the meridian. Let an instrument be so adjusted at the point A. Let some convenient station B be taken, visible at A, and let the distance AB be carefully measured. This distance is called the *base-line*. Now, by means of the telescope, adjusted to the plane of the meridian, let some point C be established on the meridian, which shall also be visible from B, and let the angles CAB and ABC be measured. We now know in the triangle ABC two angles and the included side, and can compute the distances AC and CB. We now take some suitable point D, measure the angles DCB and DBC, and knowing CB, we can obtain the distance CD. The instrument is now taken to the point C, and again established in the plane of the meridian, either by the method of Art. 36, or by *sighting back*, as it is called, to A, or by both methods combined. A third point on the meridian, E, visible from D, is then selected, the angles ECD and EDC are measured, and the distances EC and ED computed. This process is continued until the whole distance between A and G has been obtained.

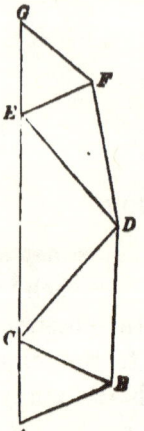

Fig. 24.

This method of measurement is called the method of *triangulation*. The base-line, AB, is purposely taken under circumstances which favor its accurate measurement, and the rest of the work consists in the determination of horizontal angles, which presents no special difficulty, and in the solution of triangles by computation.

Two things are to be noticed in reference to these triangles. The first is that in selecting the points B, C, D, &c., care must be taken so to choose them that the triangles ABC, BCD, &c., shall be nearly *equiangular;* since triangles in which there is a great inequality of the angles (*ill-conditioned* triangles, as they are called) will be much more likely to cause some error in the work.

MEASUREMENT OF THE EARTH.

The second thing to be noticed is that these triangles are really *spherical* triangles, and must therefore be solved by the formulæ of Spherical Trigonometry. If, for any reason, we are forced to take any of the points C, E, &c., off the meridian, the corresponding distances can be reduced to the meridian by appropriate formulæ.

The correctness of the result may be tested by measuring the distance GF, and comparing its measured length with that obtained by computation. The marvellous accuracy of this method of measurement is shown by the fact that in an arc of the meridian measured by the French at the close of the last century, and which was several hundred miles in length, the discrepancy between the measured and the computed length of the second base-line was less than twelve inches.

62. *Second Condition.*—The second condition requires that the angle at the centre of the earth, subtended by the arc of the meridian measured, shall be obtained. This angle is evidently the difference of latitude of the two extremities of the arc, and therefore all that is needed to satisfy this condition is that the latitude of each extremity shall be determined by appropriate observations.

Instead of determining the latitude of each place independently of the other, we may, if we choose, obtain the difference of latitude directly, by observing at each place the meridian zenith distance of the same celestial body. In Fig. 25, let A and G be the two extremities of the arc, C the centre of the earth, and S the celestial body on the meridian. If Z' is the zenith of the point A, the meridian zenith distance of S at A, reduced to the centre of the earth, is the angle $Z'CS$. In the same manner the true meridian zenith distance of S at G is the angle ZCS. The difference of these two zenith distances, or the angle ZCZ', is evidently the difference of latitude of G and A.

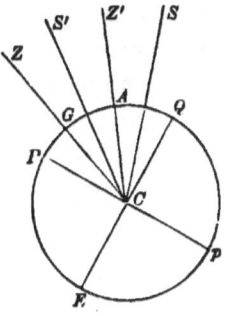

Fig. 25.

If the celestial body crosses the meridian between the two zeniths, as at S', the difference of latitude is *numerically* the sum of the two meridian zenith distances.

63. Results.—By the process above described, or by processes of a similar character, arcs of different meridians, and in different latitudes, have been carefully measured. The sum of the arcs thus measured is more than 60°, and the length of a degree of the meridian has been found to be, on the average, 69.05 miles. Multiplying this by 360, we obtain 24,858 miles for the circumference of a meridian, and dividing this circumference by π, (3.1416) we find the length of the earth's diameter to be 7912 miles.

64. *Spheroidal Form of the Earth.* —One remarkable fact is noticed when we compare the lengths of the degrees of the meridian, measured in different latitudes; and that is, that *the length of the degree is not the same at all parts of the meridian, but sensibly increases as we leave the equator.* The length of a degree at the equator is found to be 68.7 miles, whilst at the poles it is computed to be 69.4 miles. The conclusion drawn from this fact is that the figure of the earth is not rigorously that of a sphere, since a spherical form necessarily implies an absolute uniformity in the length of a degree in all parts of a great circle. In order to determine the exact geometrical figure of the earth, we must bear in mind that the curvature of a line is always proportional to the change in the direction of the tangents drawn at successive points of that line. Now, since the altitude of the elevated pole at any place is equal to the latitude of that place, it follows that an advance towards the pole of one degree in latitude is accompanied by a *depression* of one degree in the plane of the horizon. If, therefore, in order to effect a change of one degree in our latitude, we are forced to advance a greater number of miles at the pole than at the equator, we conclude that *the curvature of the meridian is less at the pole than at the equator.* Now, this same inequality in its curvature is also a peculiarity of the ellipse: and hence we infer that the form of the earth's meridians is not that of a circle, but that of an ellipse, as represented in Fig. 26. The axis of the earth, Pp, corresponds to the minor axis of an ellipse, at the extremities of which the curvature is the

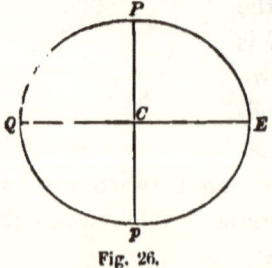

Fig. 26.

least; and the equatorial diameter of the earth, EQ, corresponds to the major axis of an ellipse, at the extremities of which the curvature is the greatest.

The form of the earth, then, is that of the solid which would be generated by the revolution of an ellipse about its minor axis, which solid is called in Geometry an *oblate spheroid*. A more common but less accurate name given to the form of the earth is that of *a sphere, flattened at the poles*.

65. *Dimensions of the Earth.*—The following are the dimensions of the earth, when its spheroidal form is taken into consideration. The determination is that of Sir G. B. Airy, the Astronomer Royal of England.

Polar diameter..........................7899.170 miles.
Equatorial diameter...................7925.648 miles.

These values are believed to be within a quarter of a mile of the true values. They differ from the results obtained by the astronomer Bessel by only about $\frac{1}{20}$th of a mile.

The *compression*, or *oblateness*, of an oblate spheroid is the ratio of the difference between the major and the minor axis of the generating ellipse to its major axis. The compression of the earth is therefore $\frac{26.478}{7925.648}$: which is about $\frac{1}{299}$th.

If a and b represent the semi-major and the semi-minor axis of the generating ellipse, the expression for the volume of the oblate spheroid is $\frac{4}{3}\pi a^2 b$. Substituting in this expression the values of a and b, we find the earth's volume to be about 260 billions of cubic miles.

66. *Density of the Earth.*—There are various methods of determining the mean density of the earth. The following is a brief summary of the method of determining it by means of the *torsion balance*. This balance consists of a slender wooden rod, supported in a horizontal position by a very fine wire at its centre. To the extremities of this rod are attached two small leaden balls. If left free to move, this horizontal rod will of course come to rest when the supporting wire is free from torsion. Two much larger leaden balls are now brought near the two suspended balls, and on opposite sides, so that the attractions of both balls may combine to twist the wire in the

same direction. The smaller balls will be sensibly attracted by the larger ones, and the horizontal rod will change the direction in which it lies. The amount of this deflection is very carefully measured, and from it is computed the attraction which the large balls exert on the small ones. But we know the attraction which the earth exerts on the small balls, it being represented by their weight: and we know also the volumes of the earth and the attracting balls. Finally, we know the density of lead: and from these data it is possible to compute the mean density of the earth.

A series of over 2000 experiments of this nature was conducted in England, in 1842, by Sir Francis Baily. The mean density of the earth, obtained from these experiments, was 5.67: the density of water being the unit. Other methods of determining the density of the earth have been employed, the main principle in each method being the comparison of the attraction exerted by the earth upon some object with that exerted by some other body, whose density can be ascertained, upon the same object. The results of these experiments do not differ materially from the results of the experiments with the torsion balance.

The volume and density of the earth being known, what is commonly called its weight can be computed. It is found to be about six sextillions of tons.

ROTATION OF THE EARTH.

67. Up to this point we have assumed the earth to be at rest, and the apparent diurnal motions of the heavenly bodies to be real motions. By careful observation of the sun, the moon, and the most conspicuous of the planets, astronomers have demonstrated that each of these bodies rotates upon a fixed axis. Analogy, therefore, points to a similar rotation of our own planet: and besides this, there are many phenomena which are inexplicable if the earth is at rest, but which are fully accounted for on the supposition that it rotates upon an axis. We will now examine the principal of these phenomena.

68. *The weight of the same body is not the same in different latitudes.* Careful experiments made in different latitudes show

that the weight of the same body is not constant at all parts of the earth's surface, but increases with the latitude. A body which weighs 194 pounds at the equator will weigh 195 pounds if taken to either pole; that is to say, the weight of any body is increased by $\frac{1}{194}$th of itself when carried from the equator to the pole. This experiment cannot be made with the ordinary balances in which bodies are weighed: since it is obvious that the same cause, whatever it may be, which affects the weight of the body will also affect that of the weights by which it is balanced, and by the same amount, so that the scales will still remain in equilibrium. If, however, we test the weight of a body (*the force*, that is to say, *with which it tends to the earth's centre*) by the effect which it has in stretching a spring, the increase of weight will be found to be as stated above.

Part of this increase of weight is due to the spheroidal form of the earth, since a body when at the pole is nearer the centre of the earth than when at the equator. The amount of increase due to this cause has been calculated to be about $\frac{1}{590}$th; hence the difference between $\frac{1}{194}$th and $\frac{1}{590}$th, which is $\frac{1}{289}$th, still remains to be accounted for. We shall now see how it is completely accounted for by the supposition that it is the effect of the centrifugal force which is induced by a rotation of the earth upon its polar axis.

69. *Centrifugal Force.*—The tendency which a body has, when revolving about any point as a centre, to recede from that centre, is called its centrifugal force. The formula for the centrifugal force may be found in any treatise on Mechanics, and is as follows:

$$f = \frac{4\pi^2 r}{t^2}$$

in which f is the centrifugal force, r the radius of the circle of revolution, and t the *periodic time*, or the time in which the revolution is performed. Now, in Fig. 27 let the earth be supposed to rotate about its polar axis, Pp, once in every sidereal day, which, as we have already seen (Art. 7), is 3m. 56s. less than the mean

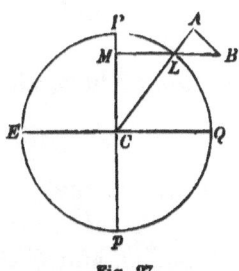

Fig. 27.

solar day, and therefore contains 86,164s. Substituting this value of t in the formula given above, and substituting for r the value of the earth's equatorial radius in feet, and computing the value of f, we shall find that the centrifugal force at the equator is .1113 feet. Now the actual force of gravity at the equator is found, by Mechanics, to be 32.09 feet. If the earth were at rest, the force of gravity at the equator would evidently be 32.09 + .1113 feet. Hence the diminution of gravity at the equator, due to centrifugal force, (in other words, the loss of weight), is equal to $\frac{.1113}{32.2013}$, or $\frac{1}{289}$th.

Since the periodic time (t in the formula) is constant for all places on the earth's surface, it is evident, from the formula, that the centrifugal force at any place L is to the centrifugal force at the equator as the radius of revolution at L, or LM, is to CQ. But we have in the figure,

$$\frac{ML}{CQ} = \frac{ML}{CL} = \cos MLC = \cos \text{Latitude}.$$

Denoting, then, the centrifugal force at the equator by C, and that at L by c, we have,—

$$c = C \cos L:$$

or *the centrifugal force varies with the cosine of the latitude.*

The centrifugal force at L acts in the direction of the radius of revolution ML. Let its amount be represented by LB, taken on LM prolonged. This force may be resolved into two forces: LA, in the direction from the centre of the earth, and AB, at right angles to LA. The force LA, being directly opposed to the attraction of the earth, has the effect of diminishing the weight of bodies at L, and may therefore be taken to represent the loss of weight at L.

Denoting the loss of weight by w, and the centrifugal force at L by c, as before, we have, from the triangle ABL,

$$w = c \cos L.$$

But we have already,—

$$c = C \cos L:$$
$$\therefore w = C \cos^2 L.$$

Now, at the equator, as is evident from the figure, the whole effect of the centrifugal force is exerted to diminish the weight of bodies, and C therefore also represents the loss of weight at the equator. We have then, finally, that *the loss of weight of a body at any latitude, due to centrifugal force, is equal to the product of* $\frac{1}{289}$*th of the weight multiplied by the square of the cosine of the latitude.*

70. Spheroidal Form of the Earth due to Centrifugal Force.— We see, then, that the supposition that the earth rotates upon its axis fully explains the observed difference in the weight of the same body in different latitudes. But this is not all: for if we assume that the particles of matter of which the earth is composed were formerly in a fluid or molten condition, and therefore free to move, the spheroidal form of the earth is itself a proof of the earth's rotation. Numerous experiments may be made to show that, for a fluid body at rest, the form of equilibrium is that of a sphere: and that, for a fluid body which rotates, the form of equilibrium is that of a spheroid, the oblateness of which increases with the velocity of rotation. Knowing the volume and the density of the earth, and assuming the time of rotation to be twenty-four sidereal hours, it is possible to calculate the form of equilibrium which a fluid mass under these conditions will assume: and this form is found to be that of a spheroid, with an oblateness very nearly identical with the known oblateness of the earth.

This tendency of a fluid mass to assume a spheroidal form under rotation may also be shown in Fig. 27. The centrifugal force LB was resolved into the two forces LA and AB, the former of which forces has already been discussed. The effect of the latter force, AB, is evidently a tendency in the particle L to move towards the equator EQ; and a similar force acting upon all the particles of matter on the earth's surface, excepting those at the poles and at the equator, will cause them all to move in the direction of the equator, and thus give a spheroidal form to the mass.

71. Trade Winds.—The *trade winds* are permanent winds which prevail in and sometimes beyond the torrid zone. These winds are northeasterly in the northern hemisphere and southeasterly in the southern hemisphere. The air within the torrid zone being, in general, subject to a greater degree of heat than the air at other portions of the earth's surface, rises, and its place is filled by air which comes in from the regions beyond the tropics. If the earth were at rest, these currents of air would manifestly have simply a northerly and a southerly direction. Now, we all know that, when we travel in any direction on a still day, or even when the wind is moving in the same direction with us, but with a less velocity, the wind seems to come from the point towards which we are going. We see from Fig. 27 that, if the earth is rotating upon its polar axis, the linear velocity of rotation decreases as the latitude increases. Hence, the air from beyond the tropics, having at the start only the linear velocity of the place which it leaves, will, as it moves towards the equator, have continually a less velocity than that of the surface over which it passes, and will seem to come from the quarter towards which those places are moving. If, then, the earth is rotating *from west to east*, these currents of air will have an apparent motion *from the east*, which motion, when compounded with the motion from the north and the south, before mentioned, will give us the northeasterly and southeasterly winds which we call the Trades.

72. *The Pendulum Experiment.**—The last and decidedly the most satisfactory proof of the earth's rotation which we shall notice, is that which comes from the apparent rotation of the plane of a freely-suspended pendulum, when made to vibrate at any point on the earth's surface except the equator.

It is an established law in Mechanics that a pendulum, freely suspended from a fixed point, always vibrates in the same plane; and also that if we give the point of support a slow movement of rotation about a vertical axis, the plane of vibration will still remain unchanged. If, for instance, we suspend a ball by a string, and, having caused it to vibrate, twist the string, the ball will rotate about the axis of the string, while the plane in which it vibrates will not be affected.

Now, let us suppose that a pendulum is suspended at the north pole, and is made to vibrate: and let us further suppose that the earth rotates from west to east, once in 24 hours. The line in which the plane of vibration intersects the plane of the horizon

---

* This is called *Foucault's experiment*. A full discussion of it is given in the American Journal of Science, 2d series, vols. XII–XIV.

PENDULUM EXPERIMENT.

will move about in the plane of the horizon, in a direction opposite to that in which the earth is rotating, and with an equal velocity, thus completing one revolution in 24 hours. In Fig. 28, let $ACBD$ be the horizon of the observer at the north pole, and let the earth rotate in the direction indicated by the arrows. Let the pendulum at P be set swinging in the direction of some diameter, AB, of the horizon. At the end of an hour, the rotation of the earth will have carried this diameter to some new position $A'B'$, at the end of the next hour to some new position $A''B''$, &c.: while the pendulum will still swing in the original direction AB. To the observer, then, unconscious of the earth's rotation, the plane of vibration, which really remains unchanged, will appear to rotate in a direction opposite to that in which the earth is rotating.

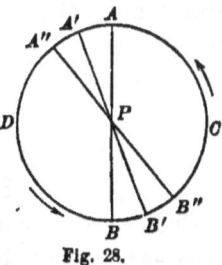

Fig. 28.

At the south pole, under the same suppositions, a similar phenomenon will be noticed, except that the plane of vibration will apparently move in the opposite direction. Thus, if at the north pole the apparent motion of the plane is like that of the hands of a clock, as we look on its face, the apparent motion at the south pole will be the opposite to this.

Again, if a pendulum is made to vibrate in the plane of a meridian at the equator, there will be *no* apparent change in the plane of vibration, since it will always coincide with the plane of the meridian, and hence the pendulum will continue to swing north and south during the entire period of the earth's rotation. The condition that the pendulum shall here swing in the plane of a meridian is entirely unnecessary, and is made only for the sake of illustration; for there will be no apparent change in the plane of vibration, whatever may be the direction in which the pendulum is made to vibrate.

The apparent rotation, then, of the plane of vibration of the pendulum is 360° in 24 hours at the poles, and nothing at the equator. At places lying between the equator and the poles, the apparent angular motion of the plane of vibration will be between these two limits; in other words, less than 360° in

24 hours. Appropriate investigations show that the apparent angular motion of the plane of vibration at any place in any interval of time is equal to the angular amount of the earth's rotation in that time, multiplied by the sine of the latitude of the place.* Thus, at Annapolis, we have for the angular motion in one hour,

$$15° \sin 38° 59' = 9° 26';$$

so that the plane of vibration will make one apparent rotation at Annapolis in 38h. 09m.

Such is the theory of the pendulum experiment. Now, numerous experiments have been made in different latitudes, and in every case an apparent rotation of the plane of vibration *from east to west* has been observed, with a rate agreeing very closely with that demanded by the theory; and the conclusion is irresistible that *the earth rotates on its polar axis, from west to east, once in every sidereal day.*

73. *Linear Velocity of Rotation.*—Taking the equatorial circumference of the earth to be 24,900 miles, we have a linear velocity of over 1000 miles an hour, and over 17 miles a minute. This is the velocity at the equator. The linear velocity at other points on the earth's surface is less than this, since the circumferences of the parallels of latitude are less than the circumference of the equator. Since the circumference of any parallel is

---

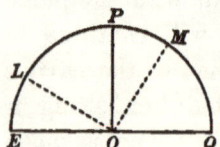

* This formula may be obtained by the principles of *the resolution of rotation*, given in treatises on Mechanics. Thus, in the figure, the rotation of the point L about the axis of the earth, PO, may be resolved into two rotations, one about the radius LO, and the other about the radius MO, drawn perpendicular to LO. If v represents the angular velocity of L about the axis PO (or 15° in one hour), and v' and v'' the angular velocities about the axes LO and MO, we have, from Mechanics,

$$v' = v \cos LOP, \text{ and } v'' = v \cos POM.$$

Now, the rotation about the axis OM will have no effect in changing the apparent position of the plane of vibration of the pendulum, since it is analogous to the case at the equator considered in the text; while the rotation about the axis LO, being analogous to the case at the pole, will produce a similar effect. The apparent angular motion, then, of the plane of vibration will be $v \cos LOP$, or $v \sin$ Lat.

… to that of the equator as the radius of the parallel is to the radius of the equator, the linear velocity will diminish as we leave the equator in the same ratio that the radii of the successive parallels diminish: in the ratio, that is, of the cosine of the latitude, as was proved in Art. 69. For instance, the cosine of 60° being ½, the linear velocity at that latitude is on'y 8½ miles a minute.

CHAPTER V.

LATITUDE. LONGITUDE.

LATITUDE.

74. THE latitude of any place on the earth's surface has been proved, in Articles 10 and 11, to be equal to either the altitude of the elevated pole or the declination of the zenith at that place. We shall now proceed to explain the principal methods by which either one or the other of these arcs may be found.

75. *First Method.*—Let Fig. 29 represent a projection of the celestial sphere on the plane of the celestial meridian, $RZHN$, of some place. HR is the celestial horizon at that place, Z the zenith, P the elevated pole, and EQ the equator. Let s represent some circumpolar star, whose declination is known, at its lower culmination. Let its meridian altitude be observed, and corrected for instrumental errors and refraction. (For all celestial bodies except the sun, the moon, and the planets, the corrections for parallax and semi-diameter will be inappreciable.) To this corrected altitude add the star's polar distance, the complement of the star's known declination. The sum is the altitude of the elevated pole, or the latitude.

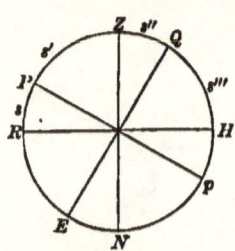

Fig. 29.

If the circumpolar star is at its upper culmination, as at s', the polar distance is to be subtracted from the corrected altitude.

If h' and h denote the corrected altitudes at the upper and the lower culmination, p' and p the corresponding polar distances, and L the latitude, we have evidently

$$L = h' - p'$$
$$L = h + p:$$

whence $\qquad L = \tfrac{1}{2} (h' + h) + \tfrac{1}{2} (p - p').$

In this formula the value of the latitude does not depend on the absolute value of either polar distance, but merely on the *change* of the polar distance between the two transits, which is usually so small as to be neglected. This method, then, is free from any error in the declination, and is used at all fixed observatories.

76. *Second Method.*—When the star is at its upper culmination, it will, in general, be more convenient to find the declination of the zenith from the meridian zenith distance of the star. Taking the star s', for instance, and denoting its meridian zenith distance by z, and its declination by d, we have

$$L = ZQ = Qs' - Zs' = d - z. \quad (a)$$

For the star s'', we have

$$L = Zs'' + Qs'' = z + d, \quad (b)$$

and for the star s'''

$$L = Zs''' - Qs''' = z - d. \quad (c)$$

From these three formulæ a general rule may be deduced, applicable to the upper culmination of every star. We notice that in the formulæ (a) and (b), where d is positive, the stars s' and s'' are on the same side of the equator with the elevated pole; that is to say, their declinations have the same name as the elevated pole; while in the formula (c) the declination has the opposite name. We also notice that in the formulæ (b) and (c), where z is positive, the stars are on the opposite side of the zenith from the elevated pole; in other words, their *bearing* has the opposite name to that of the pole: while the bearing of the star s', in the formula for which z is negative, has the same name as the elevated pole. The general rule, then, for all these stars will be the following:—If the star bears *south*, mark the zenith distance *north;* if it bears *north*, mark the zenith distance *south;* mark the declination north or south, as the star is north or south of the equator, and combine the zenith distance and the declination, thus marked, according to their names.

77. *Third Method.*—A very successful adaptation of the preceding method is made by using two stars which culminate at nearly the same time, but on opposite sides of the zenith, as s' and s'' in Fig. 29. These two stars are so selected that the difference of their zenith distances is very small, and can be mea-

sured directly by means of a micrometer. By the formulæ of the preceding article we have for s',
$$L = d - z,$$
and for s'', denoting its meridian zenith distance and declination by z' and d',
$$L = d' + z',$$
whence we have,
$$L = \tfrac{1}{2}(d + d') + \tfrac{1}{2}(z' - z).$$

The determination of the latitude is thus made free from any error in the graduations of the vertical circle, and depends only on the known declinations of the two stars, and on the *difference* of their zenith distances. Errors in the refraction are also very nearly eliminated.

This is the principle of what is called Talcott's Method, a method very commonly used by the United States Coast Survey. The instrument employed is the zenith telescope, a modification of the altitude and azimuth instrument. The two stars are so selected that the difference of their zenith distances is less than the breadth of the field of the telescope. The instrument is set in the plane of the meridian to the mean of the two zenith distances, and for the star which culminates first. When this star crosses the meridian, it is bisected by the micrometer wire, and the micrometer is read. The instrument is then turned 180° in azimuth, and the process is repeated with the second star. The difference of the zenith distances is then obtained from the difference of the two micrometer readings, and added to the half sum of the two declinations, according to the formula.

78. *Fourth Method.*—When the local time (either solar or sidereal) is known, the latitude may be obtained from altitudes which are not measured on the meridian. Let Fig. 30 be a projection of the celestial sphere on the plane of the horizon. Z is the zenith of the place, P the elevated pole, PZ the co-latitude, and S a star, whose altitude is measured. SPZ is the hour angle of the star, which can be obtained from the local time noted at the instant the altitude is observed. PS is the star's known polar distance. In the triangle

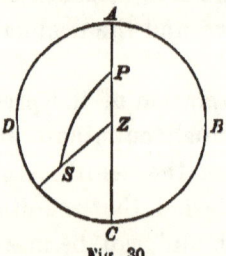
Fig. 30.

SPZ, we have the sides *ZS* and *SP*, and the angle *SPZ*, and can therefore compute the value of the co-latitude, *PZ*, by the formulæ of Spherical Trigonometry.

An analytical investigation of the formulæ by which this problem is solved shows that errors in the observed altitude and the time have the less effect upon the result the nearer the body is to the meridian.

79. *Methods of Finding the Latitude at Sea*.—The second and the fourth of the methods above described are the methods most commonly employed in finding the latitude at sea. The sun is the body which is generally used, its altitude above the sea horizon being measured with a sextant or an octant. The time of noon being approximately known, the observer begins to measure the altitude of the lower limb of the sun a few minutes before noon, and continues to measure it until the sun ceases to rise, or "dips," as it is called. The greatest altitude which the sun attains is considered to be the meridian altitude, although, rigorously speaking, it is not. The proper corrections for index-error, dip, refraction, parallax, and semi-diameter are next applied to the sextant reading, and the result is the sun's true meridian altitude, from which the latitude is obtained by the rule given in Art. 76.

When cloudy weather prevents the determination of the meridian altitude of either the sun or any other celestial body, an altitude obtained within an hour of transit, on either side of the meridian, may be used to find the latitude by the fourth method, Art. 78. Bowditch's Navigator contains special tables by which the computation, particularly when the sun is observed, may be greatly facilitated.

80. *Reduction of the Latitude*.— Owing to the spheroidal form of the earth, the vertical line at any point of the surface, as *Z'O'* in Fig. 31, which corresponds exactly with the normal drawn at that point, does not coincide with the radius of the earth, *LO*, passing through the same point, excepting at the equator

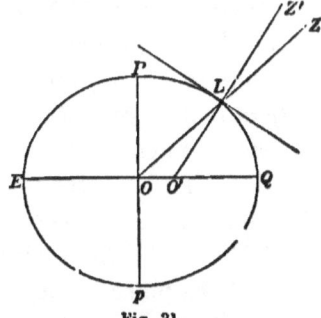

Fig. 31.

and the poles. It is necessary, then, in refined observations, to distinguish between the *geographical* zenith, Z', the point where the vertical line, when prolonged, meets the celestial sphere, and the *geocentric* zenith, Z, the point in which the radius meets the sphere. Since there are two zeniths, there are also two latitudes: $Z'O'Q$, the *geographical* latitude, and ZOQ, the *geocentric* latitude. The geographical latitude is evidently greater than the geocentric, by the angle OLO', which is called the *reduction of the latitude*. Formulæ and tables for finding this reduction are given in Chauvenet's Astronomy. It is only considered in cases where the highest accuracy in the results is required.

LONGITUDE.

81. Let Fig. 32 represent a projection of the celestial sphere on the plane of the equinoctial $ABCG$. P is the projection of

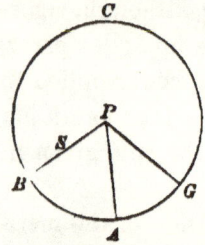

Fig. 32.

the elevated pole, and PG, PA, and PB are projections of arcs of great circles of the sphere passing through the pole. Let PG represent the projection of the meridian of Greenwich, PA that of the meridian of some other place on the earth's surface, and PB that of the circle of declination passing through some celestial body S. Then will the angle GPA represent the longitude of the meridian PA from Greenwich, GPB will represent the Greenwich hour angle of the body S, and APB will represent its hour angle from the meridian PA. The difference between these two hour angles is evidently equal to the longitude of any place on the meridian PA. The longitude, then, of any place on the earth's surface is equal to the *difference of the hour angles of the same celestial body at that place and at Greenwich, at the same absolute instant of time*. When the Greenwich hour angle is the greater of these two hour angles, reckoned always to the west, the longitude of the place is west: when it is the smaller, the longitude is east.

If PB is the hour circle passing through the sun, the longitude of the place is the difference of the *solar times* at the place and at Greenwich: if it is the hour circle passing through the vernal equinox, the longitude is the difference of the two *sidereal times*. In order, then, to determine the longitude of any place, we must be able to determine both the local and the Greenwich time (either sidereal or solar) at the same instant.

There are various methods of obtaining the local time, one of which has already been described (Art. 20). It may be noticed here that we are always able, by means of the Nautical Almanac, to convert sidereal time into solar time, or solar into sidereal (Art. 105). It remains, then, to determine the Greenwich time, either sidereal or solar, to do which several distinct methods may be employed.

82. *Greenwich Time by Chronometers.*—If a chronometer is accurately regulated to Greenwich time, that is to say, if the amount by which it is fast or slow at Greenwich on any day, and its daily gain or loss, are determined by observation, the chronometer can be carried to any other place the longitude of which is desired, and the Greenwich time which the chronometer gives can be directly compared with the time at that place. This would be a perfectly accurate method, if the rate of the chronometer remained constant during the transportation; but, in fact, the rate of a chronometer while it is carried from place to place is very rarely exactly the same that it is while the chronometer is at rest. By using several chronometers, however, and by transporting them several times in both directions between the two places, and finally by taking a mean of all the results, the error may be reduced to a very minute amount. For instance, the longitude of Cambridge, Mass., was determined by means of fifty chronometers, which were carried three times to Liverpool and back, and from them the longitude was obtained with a probable error of only $\frac{1}{8}$th of a second of time.

83. *Greenwich Time by Celestial Phenomena.*—There are certain celestial phenomena which are visible at the same absolute instant of time, at all places where they can be seen at all. Such are the beginning and the end of a lunar eclipse; the eclipses of the satellites of the planet Jupiter by that planet; the transits of these satellites across the planet's disc, and their occultations by it. The Greenwich times at which these various phenomena will occur are computed beforehand, and are pub-

lished in the Nautical Almanac. The observer, then, to obtain his longitude, has only to note the local time at which any one of these phenomena occurs, and to compare that time with the corresponding Greenwich time given in the Almanac. The difficulty of determining the exact instant at which these phenomena occur, however, diminishes to some extent the accuracy of the results. On the other hand, the times of solar eclipses and of occultations of stars by the moon, although not identical at different places, can be very accurately determined: and hence these phenomena are often employed in obtaining longitudes. (Art. 164.)

84. *Greenwich Time by Lunar Distances.*—By the *lunar distance* of a celestial body is meant its true angular distance from the centre of the moon, as it would be seen at the centre of the earth. The lunar distances of the sun, of the four brightest planets, and of nine bright stars are given in the Nautical Almanac, computed for every third hour of Greenwich solar time. An observer, then, who wishes to determine his longitude, measures the *apparent* angular distance of the moon from some one of these bodies, and also notes the local time at which the observation is made. He then finds from this apparent distance, by means of appropriate formulæ and tables, the true *geocentric* angular distance, at the time of observation, between the two bodies. He then enters the table of lunar distances in the Almanac with this distance, and finds the corresponding Greenwich time, from which, and the local time noted, he can determine his longitude.

85. *Difference of Longitude by Electric Telegraph.*—When two stations, the difference of longitude of which is desired, are connected by an electro-telegraphic wire, the difference of longitude may be determined by means of signals made at either station, and recorded at both. Suppose, for instance, there are two stations A and B, of which A is the more easterly, and that each station is provided with a clock regulated to its own local time. Let the observer at A make a signal, the time of which is recorded at each station. Let λ denote the difference of longitude of the two stations, T the local time at A at which the signal is made, and T' the corresponding time at B. Since

A is to the east of B, its time is greater at any instant than that of B. We have then, supposing the signal to be recorded simultaneously at the two stations,

$$\lambda = T - T'.$$

Experience proves, however, that the records of the signal are not exactly simultaneous, since time is required for the electric current to pass over the wire. In the example above given, then, if we denote the time required by the electric fluid to pass from A to B by x, the time recorded at B will evidently be, not T', but $T' + x$; so that the expression for the difference of longitude will be

$$\lambda' = T - T' - x.$$

Now let us suppose that instead of the signal's being made by the observer at A, it is made by the observer at B, at the time T'. The corresponding time recorded at A will not be T, but $T + x$. In this case, then, the expression for the difference of longitude will be

$$\lambda'' = T + x - T'.$$

Taking the mean of the values of λ' and λ'', we have

$$\tfrac{1}{2}(\lambda' + \lambda'') = T - T' = \lambda.$$

Any error, therefore, which is caused by the time consumed by the electric current in passing between two stations is eliminated by determining the difference of longitude by signals made at both stations, and taking the mean of the results.

86. *Difference of Longitude by "Star Signals."*—The "method of star signals" is a modification of the method described in the preceding paragraph, which is extensively used in the United States Coast Survey. The principle on which this method rests is that, since a fixed star makes one apparent revolution about the earth in exactly twenty-four sidereal hours, the difference of longitude between two meridians is equal to the interval of sidereal time in which any fixed star passes from one of these meridians to the other. The clock by which this interval of time is measured may be placed at either station, or indeed at any place which is in telegraphic communication with both stations. Two chronographs, one at each station, are connected with the wire and the clock, and upon them are

recorded, by breaks in the circuit as explained in Art. 22, the successive beats of the clock. A transit instrument is adjusted to the meridian at each station. As the star crosses the several threads of the reticule of the transit instrument at the eastern station, the observer, by means of a break-circuit key, records the instants upon both chronographs. The same process is repeated as the star crosses the wires at the western station. Now, it is evident that the elapsed time between the transits at the two meridians has been recorded upon each chronograph. Each of these values of the elapsed time is to be corrected for instrumental errors, errors of observation, and for the gain or loss of the clock in the interval; and the mean of the two values, thus corrected, is taken as the difference of longitude of the two places.

By making similar observations on several stars on the same night, by repeating the observations on subsequent nights, by exchanging observers and using different clocks, and, finally, by taking a mean of the results, a very accurate determination of the difference of longitude may be secured.

87. *Method of Finding the Longitude at Sea.*—The method of finding the longitude at sea which is usually employed is the method of Art. 82. The Greenwich time is given by chronometers regulated to Greenwich time, and the local time is obtained from the observed altitudes of celestial bodies. The sun is the body the altitude of which is most commonly used for this purpose; but altitudes of the most conspicuous of the planets and the fixed stars may also be successfully employed. Altitudes of the moon are to be avoided, except in cases where no other body is available. At the instant when the altitude of any celestial body is observed, the time shown by a watch is noted. This watch, either shortly before or after the observation, is compared with the Greenwich chronometer, and by means of this comparison the Greenwich time of the observation is obtained from the time given by the watch. The necessary corrections are applied to the sextant reading to obtain the body's true altitude. We shall then have, in the triangle PZS, Fig. 33, the side ZS, the zenith distance of the body, PS its polar distance, obtained from the Nautical Almanac, and

PZ the co-latitude of the place of observation;
the latitude being determined by some one of
the methods already given (Art. 79), and being
reduced to the time of observation by the run
of the ship given by the log. In the triangle
PZS, then, having the three sides given, we
can compute the angle *SPZ*, which is the hour
angle of the body. From this hour angle the
local time can be readily found, from which, and the Greenwich
time already obtained, the longitude may be determined.

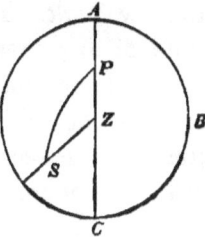

Fig. 33.

In case there is no chronometer on board, the method of
lunar distances is the only regularly available method of determining the Greenwich time. At the present day, however,
lunar distances are mainly employed as *checks* upon the chronometer, since any change in the rate of a chronometer will
cause a discrepancy between the Greenwich time shown by the
chronometer and that deduced from observation.

It can be shown, by proper methods of investigation, that
an error in the assumed latitude, or in the body's altitude,
causes the less error in the resulting hour angle the nearer
the body is to the prime vertical. It is best, then, in observing
the altitude of any celestial body for the purpose of obtaining
the local time, to observe it when the body bears nearly east or
west, provided the altitude is not so small as to be sensibly
affected by errors in the refraction. It may also be shown that
in selecting celestial bodies for observations of this character, it
is best, if the other conditions are satisfied, to take those bodies
which have the smallest declinations.

88. *Comparison of the Local Times of Different Meridians.*—
Since the local time, either solar or sidereal, is the greater
at the more easterly of any two meridians, it follows that a
watch or chronometer which is regulated to the time of any
one meridian will appear to *gain* when carried to the *west*,
and to *lose* when carried to the *east:* the amount of gain or
loss in any case being the difference of longitude, in time,
of the two meridians. A watch, for instance, which gives the
correct solar time at Boston will, even if it really is running
accurately, appear to gain nearly twelve minutes when taken

to New York. If, then, a watch which is regulated to the solar time of any meridian is carried *to the east*, the difference of longitude in time between the meridian left and that arrived at must be *added* to the reading of the watch, to obtain the time at the second meridian: if it is carried *to the west*, the difference of longitude must be *subtracted*.

CHAPTER VI.

THE SUN. THE EARTH'S ORBIT. THE SEASONS. TWILIGHT. THE ZODIACAL LIGHT.

89. *The Ecliptic.*—If a great circle on any globe is assumed to represent the celestial equator, and any point of that circle is taken to represent the vernal equinox, the *relative* positions of all bodies, the right ascension and declination of which are known, can be plotted upon this globe, and we shall have a representation of the celestial sphere. The poles of the great circle will represent the poles of the celestial sphere, and all great circles passing through these poles will represent circles of declination. We have seen, in the chapter on Astronomical Instruments, in what manner the right ascension and the declination of any celestial body can be determined at any time by observation. If we thus determine the position of the sun from day to day, and mark the corresponding points upon our celestial globe, we shall find that the sun appears to move in a great circle of the sphere from west to east, completing one revolution in this circle in 365d. 6h. 9m. 9.6s. of our ordinary solar time. This interval of time is called the *sidereal year.* The great circle in which the sun appears to move is called the *ecliptic,* and the two points in which it intersects the celestial equator are called the *vernal* and the *autumnal equinox.*

Let Fig. 34 be a representation of the celestial sphere. $EAQV$ is the equinoctial, Pp is the axis of the sphere, and P the north pole. The circle $ACVD$ represents the ecliptic, V the vernal, and A the autumnal equinox. The sun is at the vernal equinox on the 21st of March It thence moves eastward and northward, and reaches the point C,

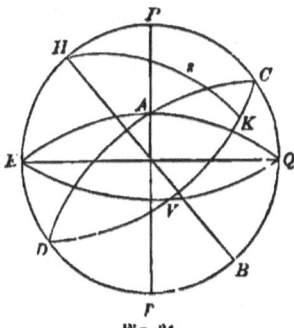

Fig. 34

where it has its greatest northern declination, on the 21st of June. This point is called the *northern summer solstice*. From this point it moves eastward and southward, passes the autumnal equinox A on the 21st of September, and reaches the point D, called the *northern winter solstice*, on the 21st of December. It thence moves towards V, which it reaches on the 21st of March.

The obliquity of the ecliptic to the equinoctial is the angle CVQ, measured by the arc CQ. This angle or arc is evidently equal to the greatest declination, either north or south, which the sun attains, and is found by observation to be about 23° 27'.

90. *Definitions.*—The *latitude* of a celestial body is its angular distance from the plane of the ecliptic, measured on a great circle passing through its poles, and called a *circle of latitude*. In Fig. 34 the arc Ks is the latitude of the body s. The *longitude* of a celestial body is the arc of the ecliptic intercepted between the vernal equinox and the circle of latitude passing through the body. Thus VK is the longitude of the body s. Longitude is properly reckoned *towards the east*.

The hour circle which passes through the solstices, the circle $DHCB$, is called the *solstitial colure*. The hour circle which passes through the equinoxes is called the *equinoctial colure*.

91. *Signs.*—The ecliptic is divided into twelve equal parts, called *signs*, which begin at the vernal equinox, and are named eastward in the following order: Aries, Taurus, Gemini, Cancer, Leo, Virgo, Libra, Scorpio, Sagittarius, Capricornus, Aquarius, Pisces. Hence the vernal equinox is called the *first point of Aries*.

The *Zodiac* is a zone or belt on the celestial sphere, extending about 9° on each side of the ecliptic.

DISTANCE OF THE SUN FROM THE EARTH.

92. *Relative Distances of the Earth and Venus from the Sun.*—It is found by observation that the *mean* value of the sun's angular semi-diameter remains constant from year to year, being always 16' 2". Since any increase or decrease in the distance of the earth from the sun will evidently be accompanied by a corresponding decrease or increase in the sun's angular semi-diameter, we conclude that the *mean* distance of

the earth from the sun is also constant from year to year. The distance of the earth from the sun is obtained by determining the sun's horizontal parallax from certain observations made upon the planet Venus. This planet revolves in a nearly circular orbit about the sun, in a plane only 3° inclined to the plane of the ecliptic. Its distance from the sun is less than that of the earth from the sun, and hence it sometimes passes between the earth and the sun, and is seen apparently moving across the sun's disc. This phenomenon is called a *transit* of Venus. As a preliminary to the determination of the earth's distance from the sun from one of these transits, it is necessary to obtain the *relative* distances of Venus and the earth from the sun. To do this, in Fig. 35 let S be the sun, E the earth, and $V V' V'' V'''$ the orbit of Venus about the sun. It is evident that the greatest angular distance (or elongation) of Venus from the sun, the greatest value, that is, of the angle VES, will occur when the line from the earth to Venus is tangent to the orbit of Venus, as represented in the figure. The orbit of Venus is not really a circle, but an ellipse, and hence the distance VS is slightly variable. So, also, is the distance SE; hence the greatest elongation is also variable, being found to lie between the limits of about 45° and 47°. Assuming its mean value to be 46°, we have in the right-angled triangle VSE,

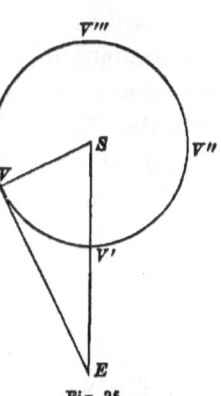

Fig. 35.

$$VS = SE \sin 46° = .72 \, SE.$$

Neglecting the inclination of the orbit of Venus to the plane of the ecliptic, we shall have, at the time of a transit, when Venus is at V',

$$V'E = .28 \, SE.$$

Hence at the time of a transit the distance of Venus from the sun is to that of Venus from the earth as about 72 to 28.*

---

* If we know the periodic time of Venus and that of the earth, the ratio of the distances of these two planets from the sun can be obtained by Kepler's Third Law (Art. 117), that "the squares of the periodic times of any two planets are proportional to the cubes of their mean distances from the sun."

93. Transit of Venus.—In Fig. 36, let S denote the centre of the sun, and $CADN$ its disc: let V be Venus, and E the centre

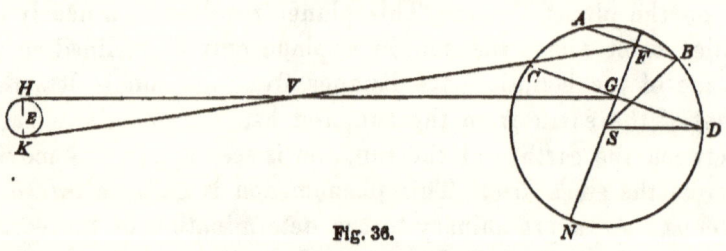

Fig. 36.

of the earth. Let HK be that diameter of the earth which is perpendicular to the plane of the ecliptic, and let an observer be supposed to be stationed at each extremity. In order to simplify the explanation, let us neglect the rotation of the earth during the observation, and suppose Venus to move in the plane of the ecliptic. To the observer at H, Venus will appear to move across the sun's disc in the chord CD, and to the observer at K, in the chord AB. Regarding VHK and VFG as similar triangles, we have, by Geometry,

$$FG : HK = GV : VH = 72 : 28$$

$$\therefore FG = \frac{18}{7} HK.$$

Again, we can obtain the angle which the line FG subtends *at the earth's centre* in the following manner. Let the observer at H note the interval of time in which the planet crosses the sun's disc in the chord CD, and the observer at K the interval in which it moves through the line AB. Since there are tables which give us the angular velocity, as seen from the earth, both of the sun and of Venus, we can deduce the angles at the earth's centre subtended by the chords FB and GD, and knowing also the angular semi-diameter of the sun, in other words, the angle at the earth's centre subtended by SB or SD, we can compute the angles at the earth's centre subtended by FS and GS, and, finally, the angle subtended by FG.

We have now determined the angle subtended by the line FG, at a distance equal to that of the earth from the sun, and also the ratio of FG to the earth's diameter. It is evidently easy to obtain from these values the angle at the sun subtended by the earth's radius, which angle is the sun's horizontal parallax, as we have already seen in Art. 54.

Although we have assumed in this discussion that the two observers are stationed at the extremities of the same diameter, it is really only necessary that they shall be at two places whose difference of latitude is large. The earth's rotation and other things which we have here neglected must be taken into consideration in the practical determination of the sun's parallax.

94. *Distance of the Earth from the Sun.*— The last two transits of Venus were in 1769 and 1874, and from observations in 1769, the sun's horizontal parallax was determined to be $8''.6$. Later observations of a different character have given a horizontal parallax of $8''.848$,* which is here used. The results of the extensive observations in 1874 cannot yet be given.

From Art. 54, we have for the distance in miles of the earth from the sun,

$$d = R \operatorname{cosec} P = 3962.8 \operatorname{cosec} 8''.848 = 92{,}400{,}000 \text{ miles.}$$

95. *Magnitude of the Sun.*— The length of the sun's radius can be at once obtained as soon as we know its distance from the earth. Thus, in Fig. 37, let S be the centre of the sun, and E that of the earth.

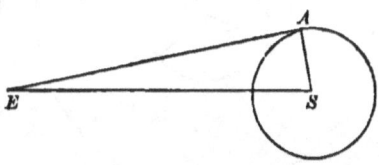

Fig. 37.

The angle AES is the apparent semidiameter of the sun, which we obtain by observation, its mean value being, as already stated, $16' 2''$. We have then, in the right-angled triangle AES,

$$SA = 92{,}400{,}000 \sin 16' 2'' = 431{,}000 \text{ miles.}$$

The sun's linear radius, then, is equal to nearly 109 of the earth's radii; and since the volumes of spheres are proportional to the cubes of their radii, the volume of the sun bears to that of the earth the enormous ratio of $1{,}286{,}000$ to 1.

By observations and calculations which will be described in the Chapter on Gravitation (see Art. 114), the mass of the sun is found to be about 327,000 times that of the earth; or about 670 times the sum of the masses of all the planets of the solar system.

---

* Determination of Professor Simon Newcomb, United States Navy. The value obtained by Leverrier is $8''.95$.

THE EARTH'S ORBIT.

96. *Revolution of the Earth about the Sun.*—Up to this point we have spoken of the apparent annual motion of the sun in the ecliptic from west to east, as though the earth were really at rest, and the sun revolved about it in its orbit. But when we take into consideration the immense mass of the sun compared with that of the earth, we are almost irresistibly led to conclude that the apparent annual revolution of the sun is the result, not of the actual revolution of the sun about the earth, but of that of the earth about the sun. Such a revolution of the earth, from west to east,* would give to the sun precisely that apparent motion in the ecliptic which has been observed. This may be seen in Fig. 38. Let S be the sun, $EE'E''$ the earth's orbit, and the outer circle $S'S''S'''$ the great circle in which the plane of the ecliptic, indefinitely extended, meets the celestial sphere. When the earth is at E, the sun will be projected in S'; when the earth is at E', the sun will be projected in S'', &c.; that is to say, while the earth moves

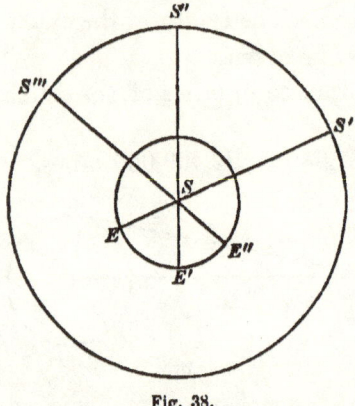

Fig. 38.

---

* Whatever the absolute motion of any celestial body moving in a circle or an ellipse may be, the appearance presented in that motion will be reversed if the spectator moves from one side of the plane in which the motion is performed to the other. Thus, the apparent daily motion to the westward of any celestial body is the same as the motion of the hands of a clock as we look upon its face, to an observer who is on the north side of the plane of the diurnal circle in which the body moves, as is seen in any latitude north of 23° 27′ in the motion of the sun; while the same westward motion presents the opposite appearance if the observer is to the south of the plane of motion, as may be seen in these latitudes in the case of the Great Bear. The appearance presented by a motion from west to east is of course the reverse of this; hence when we say that the earth or any other body moves about the sun from west to east, we mean that, *to an observer situated to the north of the plane of motion, the body appears to move in a direction opposite to that in which the hands of a clock move.*

about the sun in the direction $EE'E''$, the sun will apparently move about the earth in the same direction, $S'S''S'''$.

Against this theory, then, of the earth's revolution there is nothing to urge; and analogy gives us a strong argument in favor of it. Almost every celestial body in which any motion at all can be detected is found to be revolving about some other body, larger than itself. The moon revolves about the earth; the satellites of the planets revolve about the planets; and the planets themselves, some of which are much larger than the earth, and at a much greater distance from the sun, revolve about the sun. Henceforward, then, we shall include the earth in the list of planets, and consider the sidereal year to be the interval of time in which the earth makes one complete revolution about the sun.

97. *Linear Velocity of the Earth in its Orbit.*—The number of miles in the circumference of the earth's orbit, considered as a circle, is obtained by multiplying the radius of the orbit by 2π. If we then divide this product by the number of seconds in a year, we shall have, in the quotient, the number of miles through which the earth moves about the sun in a second of time. It will be found to be about 18.4 miles.

98. *Elliptical Form of the Earth's Orbit.*—Although, as has already been stated, the mean value of the sun's angular semi-diameter remains constant from year to year, careful measurements of the semi-diameter show that it varies in magnitude *during the year*, being greatest about the first of January, and least about the first of July. The evident conclusion from this fact is that the distance between the earth and the sun also varies during the year, being greatest when the sun's semi-diameter is the least, and least when it is the greatest. The truth of this conclusion may be seen in Fig. 37, in which we have

$$\sin AES = \frac{AS}{ES}.$$

As AS of course remains constant, ES will vary inversely as sin AES, or since the sines of small angles are proportional to the angles themselves, inversely as the angle AES itself. The greatest angular semi-diameter of the sun is 16' 17".8, the least is 15′ 45″.5: hence the ratio of the greatest to the least distance is that of 16′ 17″.8 to 15 45″.5, or of 1.034 to 1.

Let us now assume any line, SA in Fig. 39, for instance, as our unit of measure, and prolong it until SH is to SA as 1.034 is to 1. Then if S denotes the sun, SA and SH will represent the relative distances of the earth from the sun on about the first of January and the first of July. On certain days throughout the year, let the advance of the sun in longitude since the time when the earth was at A be determined, and let the angular semi-diameter of the sun on each of these days be measured. Lay off the angles ASB, ASC, &c., equal to these advances in longitude. Since, as may readily be seen in Fig. 38, the apparent advance of the sun in longitude is caused by the advance of the earth in its orbit, and is equal to it, the angles ASB, ASC, &c., will represent the angular distances of the earth from the point A on the days when the different observations were made. Let us next take the lines SB, SC, &c., of such lengths that each line may be to SA in the inverse ratio of the corresponding semi-diameters. If, then, we draw a line through the points A, B, C, &c., we shall have a representation of the orbit of the earth about the sun. The curve is found to be an ellipse, the sun being at one of the foci. The point A, where the earth is nearest to the sun, is called the *perihelion*, the point H, the *aphelion;* and the angular distance of the earth from its perihelion is called its *anomaly*.

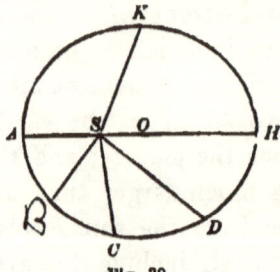

Fig. 39.

The eccentricity of the ellipse, or if O is the centre of the ellipse, the ratio of OS to OA, is evidently equal to about $\frac{0.17}{1.017}$, or $\frac{1}{60}$th. A more accurate value of it is .0167917. This eccentricity is at present subject to a diminution of .000041 a century; but Leverrier, a French astronomer, has proved that after the eccentricity has diminished to a certain point it will begin to increase again.

THE SEASONS.

99. The change of seasons on the earth is caused by the

inequality of the days and nights, and this inequality is a result of the inclination of the plane of the equinoctial to that of the ecliptic. The relative positions of the sun and the earth at different parts of the year are represented in Fig. 40. S represents

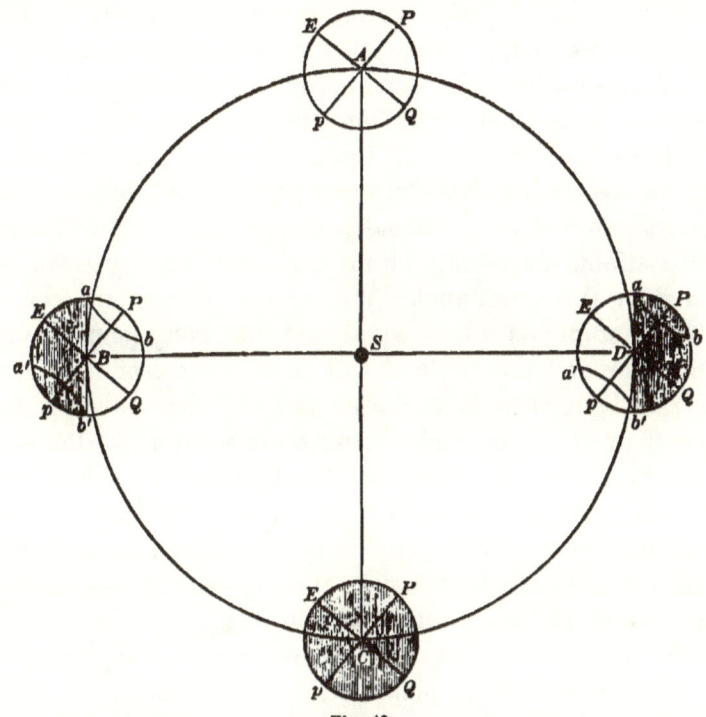

Fig. 40.

the sun, and $ABCD$ the orbit of the earth. Pp is the axis of rotation of the earth, and EQ the equator. The plane of this equator is supposed to intersect the plane of the ecliptic in the line of equinoxes AC. Since, as we have already seen, the sun appears to be on this line on the 21st of March and the 21st of September, the earth itself must also be on this line at the same time. Suppose, then, the earth to be at A on the 21st of March. The sun will evidently lie in the direction AS, and will be projected on the celestial sphere at the vernal equinox. Now, since a line which is perpendicular to a plane is perpendicular to every line in that plane which is drawn to meet it, the axis Pp at A, being perpendicular to the equator, is also perpendicular to the line AS, which is common to both the plane of the equator and the plane of the ecliptic. Half of each parallel of latitude on the earth will therefore lie in light and half in darkness; and hence, as the earth rotates on the axis Pp, every point on its surface will describe half of its diurnal course in light and half in darkness: in other words, day and night will be equal over the whole earth. Since the direction of the axis of rotation remains unchanged, the same condition of things will occur when the earth is at C, on the 21st of September. Let the earth be at B on the 21st of June. Here we see that, as the earth rotates on its axis Pp, every point on its surface within the circle ab will lie continually in the light, and will hence have continual day, while within the corresponding circle $a'b'$ the night will be continual. We see also that at the equator the days and nights will be equal, and that every point between the equator and the circle ab will describe more of its diurnal course in light than in darkness, and will thus have its days longer than its nights; while between the equator and the circle $a'b'$ the nights will be longer than the days. Similar phenomena will occur when the earth is at D, on the 21st of December, except only that it will then be the southern hemisphere in which the days are longer than the nights, and the southern pole at which the sun is continually visible.

Such, then, is the inequality of the days and nights caused by the inclination of the plane of the equinoctial to that of the ecliptic. As the sun apparently moves from either equinox, the inequality of day and night continually increases, reaches its maximum when the sun arrives at either solstice, and then continually decreases as the sun moves on to the equinox: the day being longer than the night in that hemisphere which is on the same side of the equator with the sun. Now, any point on the earth's surface receives heat during the day and radiates it during the night: and hence, when the days are longer than the nights, the amount of heat received is greater than the amount radiated, and the temperature increases; while, on the contrary, when the days are shorter than the nights, the temperature decreases: and thus is brought about the change of seasons on the earth.

Another fact, depending on the same cause, and tending to

the same result, must also be taken into consideration; and that is that the temperature at any place depends on the obliquity of the sun's rays: on the altitude, in other words, which the sun attains at noon. Now we have, from Art. 76,

$$z = L - d:$$

from which we see that, the latitude remaining constant, the sun attains the greater altitude, the greater its declination when it has the same name as the latitude, and the less its declination when it has the opposite name: so that the nearest approach to verticality in the sun's rays will occur at the same time that the day is the longest. An exception, however, must be noticed to this general rule, in the case of places within the tropics: since at these places, as may be seen from the formula, the sun passes through the zenith when its declination is equal to the latitude, and has the same name.

100. *Effect of the Ellipticity of the Earth's Orbit on the Change of Seasons.*—The elliptic form of the earth's orbit has very little to do with the change of seasons. For although the earth is nearer to the sun on the 1st of January than on the 1st of July, yet its angular velocity at that time is found by observation to be greater, and to vary throughout the whole orbit inversely as the square of the distance. Now it may readily be shown that the amount of heat received by the earth at different parts of its orbit also varies, other things being equal, inversely as the square of the distance: so that equal amounts of heat are received by the earth in passing through equal angles of its orbit, in whatever part of its orbit those angles may be situated. Still, although the change in distance does not materially affect the annual change of seasons, it does affect the *relative intensities* of the northern and the southern summer. The southern summer takes place when the earth's distance is only about $\frac{29}{30}$ths of what it is at the time of the northern summer: hence, the intensity at the former period will be to that at the latter in the ratio of about $(\frac{30}{29})^2$ to 1, or about $1\frac{9}{14}$ to 1: in other words, the intensity of the heat of the southern summer will be $\frac{1}{14}$th greater than that of the heat of the northern summer.

TWILIGHT.

101. If the earth's atmosphere did not contain particles of dust and vapor, which serve to reflect the rays of light, the transition from day to night would be instantaneous, and the intermediate phenomenon of twilight would have no existence.

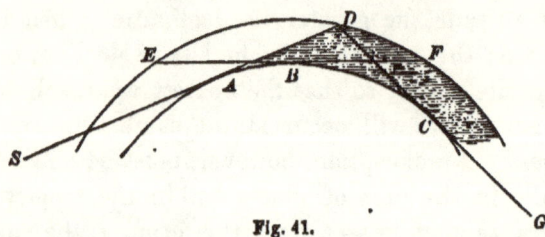

Fig. 41.

This phenomenon is explained in Fig. 41, in which ABC represents a portion of the earth's surface, and EDF a portion of the atmosphere. Let the sun be supposed to lie in the direction AS, and to be in the horizon of the place A. All of the atmosphere which lies above the horizontal plane SD will then receive the direct rays of the sun, and A will receive twilight from the whole sky. The point B will, on the contrary, be illuminated only by the smaller portion of the atmosphere included within the planes EB and AD and the curved surface ED; and at the point C the twilight will have wholly ceased. Strictly speaking, the lines AS, BE, &c., should be slightly curved, owing to the effects of refraction, but the omission involves no change in the explanation.

It is computed that twilight ceases when the sun is about 18° below the horizon, measured on a vertical circle. The more nearly perpendicular to the horizon is the diurnal circle in which the sun appears to move, the more rapid will be the sun's descent below the horizon; hence, the length of twilight diminishes as we approach the equator and increases as we recede from it. Furthermore, we see in Fig. 2 that the greater the declination of the sun, the smaller is the apparent diurnal circle in which it moves, and the greater will be the length of time required for the sun to reach the depression of 18° below the horizon. The shortest twilight, therefore, occurs at places on the equator, when the sun is on the equinoctial, and its length is then 1h. 12m. Near the poles the length of twilight is at times very great. Dr. Hayes, in his last expedition towards the North

APPEARANCE OF THE SUN. 95

Pole, wintered at latitude 78° 18' N., so far above the circle *ab*, Fig. 40, that the sun was continually below the horizon from the middle of October to the middle of February; but at the beginning and the end of this interval twilight lasted for about nine hours. At the poles twilight lasts nearly a month and a half.

GENERAL DESCRIPTION OF THE SUN.

102. When the sun is observed with a telescope, *spots* are noticed upon its surface. These spots appear to cross the sun's disc from east to west, and with different rates, the rate of motion of spots at the sun's equator being the greatest. We therefore conclude that these spots have not only an *apparent* motion, caused by the sun's rotation, but also a *proper* motion of their own. By appropriate investigation of these motions, it is found that the sun rotates from west to east upon a fixed axis, in a plane inclined at an angle of about 7° to the plane of the ecliptic. The period of this rotation is about 25 days.

Much uncertainty exists as to the nature of these spots. Until recently, it was held that the sun is surrounded by two atmospheres, of which only the outer one (called the *photosphere*) is luminous, and that the spots are rents in these atmospheres through which the solid body of the sun is seen. These spots are for the most part confined to a zone, extending about 35° on each side of the sun's equator. They differ widely in duration, sometimes lasting for several months, and sometimes disappearing in the course of a few hours. They are sometimes of an immense size. One was seen in 1843, with a diameter of nearly 75,000 miles: it remained in sight for a week, and was visible to the naked eye. In 1858, a much larger one was seen, its diameter being over 140,000 miles. As a general thing, each dark spot, or *umbra*, as it is called, has within it a still darker point, called the *nucleus*, and is surrounded by a fringe of a lighter shade, called the *penumbra*. Sometimes several spots are inclosed by the same penumbra; and occasionally spots are seen without any penumbra at all.

On the theory of two atmospheres, the existence of the penumbra is explained by supposing the aperture in the outer and luminous stratum to be wider than that in the inner one, and that portions of the inner stratum, being subjected to a strong light from above, are rendered visible: the umbra itself being, as already remarked, the solid body of the sun seen through both strata. According to Mr. J. N. Lockyer, "sun-spots are cavities or hollows eaten into the photosphere, and these different shades [the penumbra, umbra, and nucleus] represent different depths." [See Note, page 99.]

One very curious and interesting discovery in relation to these spots is that of a periodicity in their number. This discovery was made by Schwabe, of Dessau, whose researches and observations on this subject covered a period of more than twenty-five years. The number of groups of spots which he observed in a year varied from 33 to 333, the average being not far from 150.* He found the period from one maximum to another to be about ten years. Professor Wolf, of Zurich, after tabulating all the observations of spots since 1611, decided that the period varied from eight to sixteen years, its mean value being about eleven years. Recent investigations show that this periodicity is in some way connected with the action of the planets, of Jupiter and Venus particularly, upon the sun's photosphere. It is a curious fact that magnetic storms and the phenomenon called Aurora or Northern Lights have a similar period, and are most frequent and most striking when the number of the solar spots is the greatest.

Still other phenomena which are seen upon the sun's disc are the *faculæ*, which are streaks of light seen for the most part in the region of the spots, and which are undoubtedly elevations or ridges in the photosphere: and the *luculi*, which are specks of light scattered over the sun's disc, giving it an appearance not unlike that of the skin of an orange, though relatively much less rough. The cause of these luculi is unknown.

At the time of a total eclipse of the sun by the moon, the disc of the sun is observed to be surrounded by a ring or halo of light, which is called the *corona*. The breadth of this corona is more than equal to the diameter of the sun. Many theories have been advanced to explain this phenomenon, one of which is that it is due to the existence of still another atmosphere, exterior to the

---

* A table of Schwabe's observations is given in the Appendix.

photosphere. Another theory is that this corona consists of streams of luminous matter, radiating in all directions from the sun. *Rose-colored protuberances*, sometimes called *red flames*, are also seen, which are usually of a conical shape, and are sometimes of great height. In the total eclipse of August 17th, 1868, one was observed with an apparent altitude of 3', corresponding to a height of about 80,000 miles. These protuberances were formerly supposed to be similar in character to our terrestrial clouds; but Dr. Jannsen, the chief of the French expedition sent out to the East to observe the total eclipse of August, 1868, examined their light with the spectroscope, and found them to be masses of incandescent gas, of which the greater part was hydrogen. Dr. Jannsen also made the interesting discovery that these protuberances can be examined at any time, without waiting for the rare opportunity afforded by a total eclipse. He observed them for several successive days, and found that great changes took place in their form and size. Mr. Lockyer, of England, who has since examined them, pronounces them to be merely local accumulations of a gaseous envelope completely surrounding the sun: the spectrum peculiar to these protuberances appearing at all parts of the disc.

It has already been stated (Art. 47) that the spectroscope enables us to establish the existence of certain chemical substances in the sun, by a comparison of the spectra of these substances with that of the sun; or, more precisely, by a comparison of the lines, bright or dark, by which these different spectra are distinguished. The number of the parallel dark lines in the solar spectrum which have been detected and mapped exceeds 3000; and careful examination also shows that some of these are double. Some of the more prominent of these lines have received the names of the first letters of the alphabet; D, for example, is a very noticeable double line in the orange of the spectrum. When certain chemical substances are evaporated, either in a flame or by the electric current, the spectra which they form are also characterized by lines, which, however, are not dark, but *bright*. If, for instance, sodium is introduced into a flame, its incandescent vapor produces a spectrum which is characterized by a brilliant double band of yellow; and it is especially noticeable that this yellow band coincides exactly in position with the dark line D of the solar spectrum. In the same way the spectrum of zinc is found to contain bands of red and blue: that of copper contains bands of green: and, in general, the spectrum of each metal contains certain bright bands or lines, peculiar to itself, and readily recognized. We may therefore conclude that *an incandescent gas or vapor emits rays of a certain refrangibility and color, and those rays only.*

Again, it is proved by experiment that if a ray of white light be allowed to pass through an incandescent vapor, *the vapor will absorb precisely those rays which it can itself emit.* If, for instance, a continuous spectrum be formed by a ray of intense white light from any source, and if the vapor of sodium be introduced in the path of this ray, between the prism and the source of light, a dark band will appear in the spectrum, identical in position with the bright yellow band which we have already noticed in the spectrum of sodium, and which we found to be identical in position with the dark line D of the solar spectrum.

We are now ready to apply the principles established by these experiments to the case of the sun. The sun is, as we saw above, a sphere surrounded by a vaporous envelope. This sphere would of itself emit all kinds of rays, and therefore give a continuous spectrum; but the photosphere which surrounds it absorbs those of the sun's rays which it can itself emit. The dark line D of the solar spectrum shows, as in the experiment above described, that sodium has been introduced in the path of the sun's rays: in other words, that sodium is in the sun's photosphere. In the same way, Professor Kirchhoff, to whom we owe this remarkable discovery, has established the existence in the photosphere of iron, calcium, magnesium, chromium, and other metals. In the case of iron, more than 450 bright lines have been detected in its spectrum: and for every one of these lines there is a corresponding dark line in the solar spectrum.

We also see, from the preceding experiments, how the presence of bright lines in the spectrum of the rose-colored protuberances could prove to Dr. Janssen that these protuberances were not masses of clouds, reflecting the light of the sun, but masses of incandescent vapor. We shall see another instance of the same description when we come to examine some of the nebulæ.

THE ZODIACAL LIGHT.

103. At certain seasons of the year a faint nebulous light, not unlike the tail of a comet, is seen in the west after twilight has ended, or in the east before it has begun. This is called the *Zodiacal Light*. Its general shape is nearly that of a cone, the base of which is turned towards the sun. The breadth of the base varies from 8° to 30° of angular magnitude. The apex of the cone is sometimes more than 90° to the rear or in advance of the sun. According to Humboldt, it is almost always visible, at the times above stated, within the tropics: in our latitudes it is seen to the best advantage in the evening near the first of March, and in the morning near the middle of October.

Of the many theories proposed to account for the zodiacal light, the one which seems to be most widely accepted is that it consists of a ring or zone of rare nebulous matter encircling the sun, which reaches as far as the earth, and perhaps extends beyond it. According to another theory, it is a belt of meteoric bodies surrounding the sun. A very interesting and valuable series of observations upon the Zodiacal Light was made by Chaplain Jones, United States Navy, in the years 1853-5, in latitudes ranging from 41° N. to 53° S. The conclusion which he drew from his observations was that the light was a nebulous ring encircling the earth, and lying within the orbit of the moon.

NOTE.—The surface of the sun is now the object of assiduous observation and study. The shining surface of the sun, whence come our light and heat, is called *the photosphere*. Outside of this is *the chromosphere*, to which the *red flames* belong, and which is largely composed of hydrogen and the vapors of metals. Outside of the chromosphere lies *the corona*. Modern research has disproved much that was formerly believed concerning the physical constitution of the sun and its envelope; but it has by no means established what that constitution really is.

CHAPTER VII.

SIDEREAL AND SOLAR TIME. THE EQUATION OF TIME. THE CALENDAR.

104. *Sidereal and Solar Days.*— It is important to distinguish between the apparent *annual* motion of the sun in the ecliptic, from west to east, and the apparent *diurnal* motion towards the west, which the rotation of the earth gives to all celestial bodies and points. A sidereal day is the interval of time between two successive transits of the vernal equinox over the same branch of the meridian. A solar day is the interval between two similar transits of the sun. But the continuous motion of the sun towards the east causes it to appear to move more slowly towards the west than the vernal equinox moves. The solar day is therefore longer than the sidereal day, the average amount of the difference being 3m. 55.5s. And furthermore, in the interval of time in which the sun makes one complete revolution in the ecliptic, the number of *daily* revolutions which it appears to make about the earth will be less by one than the number of daily revolutions made by the equinox. The sidereal year, then (Art. 89), which contains 365d. 6h. 9m. 9.6s. of solar time, contains 366d. 6h. 9m. 9.6s. of sidereal time. [See Note, page 154.]

105. *Relation of Sidereal and Solar Times.*—Since the sidereal day is shorter than the solar day (and, consequently, the sidereal hour, minute, &c., than the solar hour, minute, &c.), it is evident that any given interval of time will contain more units of sidereal than of solar time. The relative values of the sidereal and the solar days, hours, &c., are obtained as follows:—
We have from the preceding article,

366.25636 sidereal days = 365.25636 solar days:
one sidereal day = 0.99727 solar day,
one sidereal hour = 0.99727 solar hour, &c.

Having, therefore, an interval of time expressed in either solar or sidereal units, we may easily express the same interval in units of the other denomination. This is called the *conversion* of a solar into a sidereal interval, and the reverse: and tables for facilitating this conversion are given in the Nautical Almanac.

Again, knowing the sidereal time at any instant, the hour-angle, that is to say, of the vernal equinox, the corresponding solar time, or the hour-angle of the sun, is readily obtained by subtracting from the sidereal time the sun's right ascension. This is indeed a corollary of the theorem proved in Art. 9, from which we see that the sum of the sun's right ascension (which can always be found in the Nautical Almanac), and its hour-angle, is the sidereal time. Either of these times, then, may be converted into the other.

THE EQUATION OF TIME.

106. *Inequality of Solar Days.*—Observation shows that the length of the solar day is not a constant quantity, but varies at different seasons of the year, and, indeed, from day to day. A distinction must therefore be made between the *apparent* or *actual* solar day, and the *mean* solar day, which is the mean of all the apparent solar days of the year. A uniform measure of time may be obtained from the apparent diurnal motion with reference to our meridian of a fixed celestial body or point. It may also be obtained from the apparent diurnal motion of a celestial body which changes its position in the heavens, provided that two conditions are satisfied; first, the plane in which the body moves must be perpendicular to the plane of the meridian: and second, its motion in that plane must be uniform. Both these conditions are so very nearly satisfied by the motion of the vernal equinox, that any two sidereal days may be considered to be sensibly equal to each other; but neither condition is satisfied by the motion of the sun. It moves in the ecliptic, the plane of which is not, in general, perpendicular to the plane of the meridian: and its motion in this plane is not uniform. We have, therefore, two causes of the inequality of the solar days, the effect of each of which we will now proceed to examine.

107. *Irregular Advance of the Sun in the Ecliptic.*—Observation shows that the sun's motion in longitude is not uniform. The *mean* daily motion is, of course, obtained by dividing 360° by the number of days and parts of a day in a year, and is 59' 8."3. But the daily motion about the first of January is 61' 10", while about the first of July it is only 57' 12". In Fig. 42, let the circle $AM'M''$ represent the apparent orbit of the sun in the ecliptic about the earth E, and let the sun be supposed to be at the point A where its daily motion is the greatest, on the first of January. Let us also suppose a fictitious sun (which we will call the *first mean sun*) to move in the ecliptic with the uniform rate of 59' 8".3 daily, and to be at the point A at the same time that the true sun is there. On the next day the mean sun will have moved eastward to some point M, while the true sun, whose daily motion is at this time greater than that of the mean sun, will be found at some point T, to the east of M. The true sun will continue to gain on the mean sun for about three months, at the end of which time the mean sun will begin to gain on the true sun, and will finally overtake it at the point B, on the first of July. During the second half of the year the mean sun will be to the east of the true sun, and at the end of the year the two suns will again be together at A.

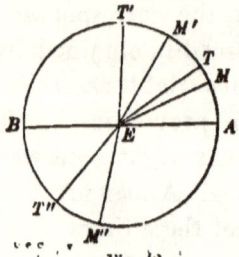

Fig. 42.

The angular distance between the two suns, represented in the figure by the angles TEM, $T'EM'$, &c., is called the *Equation of the Centre*. It is evidently additive to the mean longitude of the sun while it is moving from A to B, and subtractive from it while it is moving from B to A. Its greatest value is about 8 minutes of time.

Since the rotation of the earth gives to both these bodies a common daily motion to the west, it is plain that from January to July the mean sun will cross the meridian before the true sun, and that from July to January the true sun will cross the meridian before the mean sun.

108. Obliquity of the Ecliptic to the Meridian.—Even if the sun's motion in the ecliptic were uniform, equal advances of the sun in longitude would not be accompanied by equal advances in right ascension, in consequence of the obliquity of the ecliptic to the meridian. The truth of this may be seen in Fig. 43. Let this figure represent the projection of the celestial sphere on the plane of the equinoctial colure $PApH$. A and H are the equinoxes, P and p the celestial poles, AeH the equinoctial, and AEH the ecliptic. Let the ecliptic be divided into equal arcs, AB, BC, &c., and through the points of division, B, C, &c., let hour-circles be drawn, meeting the equinoctial in

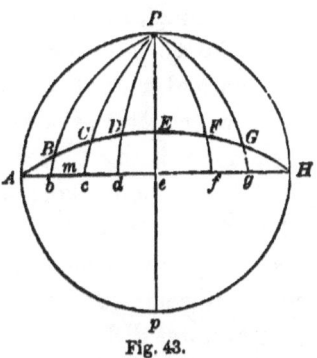
Fig. 43.

the points b, c, &c. Now, since all great circles bisect each other, AEH is equal to AeH, and if Pep is the projection of an hour-circle perpendicular to the circle $PApH$, AE and Ae are quadrants, and equal. The angle PBC is evidently greater than PAB, PCD is greater than PBC, &c.: in other words, the equal arcs AB, BC, &c., are *differently inclined* to the equinoctial. The effect of this is that the equinoctial is divided into unequal parts by the hour-circles Pb, Pc, &c., bc being greater than Ab, cd than bc, &c. It is to be noticed, further, that the points B and b, being on the same hour-circle, will be on the meridian at the same instant of time: and the same is true of C and c, D and d, &c.

Now, if A is the vernal equinox, the first mean sun, moving in the ecliptic with the constant daily rate of $59'\ 8''.3$, will pass through that point on the 21st of March. Let another fictitious sun (called the *second mean sun*) leave the point A at the same time, and move in the equinoctial with the same uniform daily rate. Since BAb is a right-angled triangle, Ab is less than AB. Hence, when the first mean sun reaches B, the second mean sun will be at some point m, to the east of b: when the first mean sun is at C, the second mean sun will be to the east of c, &c.: and the second mean sun will continue to lie to the east of the first mean sun until the 21st of June (the summer solstice), when both suns will be at the points E and e at the same instant of time, and will therefore come to the meridian together. In the second quadrant, the second mean sun will lie to the west of the first mean sun, and both suns will reach H, the autumnal equinox, on the 21st of September. The relative positions in the third and the fourth quadrant will be identical with those in the first and the second.

From the 21st of March, then, to the 21st of June, the second mean sun, being to the east of the first mean sun, will come *later* to the meridian; and the same will also be true from the 21st of September to the 21st of December. In the two other similar periods the case will be reversed, and the second mean sun will come *earlier* to the meridian than the first mean sun. The greatest difference of the hour-angles of these two mean suns is about 10 minutes of time.

109. *Equation of Time.* —It is by means of these two fictitious suns that we are able to obtain a uniform measure of time from the irregular advance of the sun in the ecliptic. The second mean sun satisfies the two conditions stated in Art. 106, and therefore its hour-angle is perfectly uniform in its increase. This hour-angle is the *mean solar time* of our ordinary watches and clocks. The hour-angle of the true sun is called the *apparent solar time:* and the difference at any instant between the apparent and the mean solar time is called the *equation of time.*

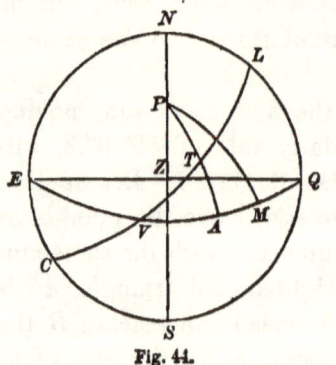

Fig. 44.

Let Fig. 44 be a projection of the celestial sphere on the plane of the horizon. Z is the zenith, P the pole, EVQ the equinoctial, CL the ecliptic, and V the vernal equinox. Let T be the position of the true sun in the ecliptic, and M that of second mean sun in the equinoctial. The angle TPM is evidently the equation of time. This angle is measured by the arc AM, or $VM - VA$: the difference, that is, of the right ascensions of the true and the second mean sun. But since the angular advance of the second mean sun in the equinoctial is, by hypothesis, as shown in the previous article, equal to the angular advance of the first mean sun in the ecliptic, it follows that the right ascension of the second mean sun is always equal to the longitude of the first mean sun, or, as it is usually called, the true sun's mean longitude. The equation of time, then, is *the difference of the sun's true right ascension and mean longitude;* and thus computed is given in the Nautical Almanac for each day in the year. It reduces to zero four times in the year, and passes through four maxima, ranging in value from 4 minutes to 16 minutes.

110. *Astronomical and Civil Time.*—The mean solar day is considered by astronomers to begin at mean noon, when the second mean sun (usually called simply the mean sun) is at its upper culmination. The hours are reckoned from 0h. to 24h. The mean solar day, so considered, is called the *astronomical day.*

The civil day begins at midnight, twelve hours before the astronomical day, and is divided into two parts of twelve hours each, called A.M. and P.M.

We must, therefore, carefully distinguish between any given civil time and the corresponding astronomical time. For instance, January 3d, 8 A.M., in civil time, is the same as January 2d, 20h., in astronomical time.

THE CALENDAR.

111. Owing to causes which will be explained further on, the position of the vernal equinox is not absolutely stationary, but moves *westward* along the ecliptic, with an annual rate of about 50″.2. The sun, then, moving eastward from the equinox, will reach it again before it has made one complete sidereal revolution about the earth. This interval of time in which the sun moves from and returns to the equinox is called a *tropical* year, and consists of 365d. 5h. 48m. 47.8s. The *Julian* Calendar was established by Julius Cæsar, 44 B.C., and by it one day was inserted in every fourth year. This was the same thing as assuming that the length of the tropical year was 365d. 6h., instead of the value given above, thus introducing an accumulative error of 11m. 12s. every year. This calendar was

adopted by the Church in 325 A.D., at the Council of Nice, and the vernal equinox then fell on the 21st of March. In 1582, the annual error of 11m. 12s. caused the vernal equinox to fall on the 11th of March, instead of the 21st. Pope Gregory XIII. therefore ordered that ten days should be omitted from the year 1582, and thus brought the vernal equinox back again to the 21st of March. Furthermore, since the error of 11m. 12s. a year amounted to very nearly three days in 400 years, it was decided to leave out three of the inserted days (called *intercalary* days) every 400 years, and to make this omission in those years which were not exactly divisible by 400. Thus of the years 1700, 1800, 1900, 2000, all of which are leap years according to the Julian calendar, only the last is a leap year according to the *reformed* or *Gregorian* calendar. By this calendar the annual error is only 24 seconds, and will not amount to a day in much less than 4000 years.

This reformed calendar was not adopted by England until 1752, when eleven days were omitted from the calendar. The two calendars are now often called the *old style* and the *new style*. For instance, April 26th, O.S., is the same as May 8th, N.S. In Russia the old style is still retained, though it is customary to give both dates; as 1868, $\frac{\text{Jan. 23}}{\text{Feb. 4}}$. All other Christian countries have adopted the new style.

CHAPTER VIII.

LAW OF UNIVERSAL GRAVITATION. PERTURBATIONS IN THE EARTH'S ORBIT. ABERRATION.

112. *The Law of Universal Gravitation.*—The earth, as we have seen in Chapter VI., revolves about the sun in an elliptical orbit, with a linear velocity of eighteen miles a second. At every point of its orbit the centrifugal force induced by this revolution must create in the earth a tendency to leave its orbit, and to go off in the direction of a tangent to the orbit at that point. To counteract this centrifugal force, there must constantly exist a centripetal force, by which the earth is at every instant deflected from this rectilinear path which it tends to follow, and is drawn towards the sun; and in order that the orbit of the earth may remain unchanged in form,—as observation shows that it does remain,—these two forces must be in constant equilibrium. Admitting, then, the existence of such a centripetal force, it remains to determine the nature of the force, and the laws under which it acts.

The force is believed to be identical in nature with that force which causes all bodies, free to move, to tend towards the earth's centre, and which we call the force of gravity. At whatever height above the surface of the earth the experiment may be made, this attractive force of the earth is found to exist; and there is no good reason for assuming any finite limit beyond which this force, however much its effects may be lessened by other and opposing forces, does not have at least a theoretic existence. And, furthermore, as the sun and the other heavenly bodies are all masses of matter like the earth, there is every reason for concluding that they too, as well as the earth, possess this power of attracting other bodies towards their centres. Nor is this attractive power a characteristic of large bodies alone: for we have already seen in the experiment with the torsion balance, described in Art. 66, that small globes of lead exert a sensible attraction upon still smaller globes. We may therefore assume that what is true of each of these masses, large and small, as a whole, is no less true of the separate particles of which it is composed, and that every particle of matter in the universe has an attractive power upon every other particle.

In order to determine the laws under which this attractive power is exerted, we have only to assume that the laws which are shown by experiment to obtain at the earth's surface hold equally good throughout the universe; so that whatever the masses of bodies may be, or whatever the distances by which they are separated from each other, the forces with which any two bodies attract a third will be directly proportional to the masses of the two attracting bodies, and inversely proportional to the squares of their distances from the third body.

This, then, is Newton's Law of Universal Gravitation. *Every particle of matter in the universe attracts every other particle, with a force directly proportional to the mass of the attracting particle, and inversely proportional to the square of the distance between the particles.* In applying this general law to the particles which compose the masses of the heavenly bodies, Newton has demonstrated that the attraction exerted by a sphere is precisely what it would be if all the particles in the sphere were collected at its centre, and constituted one particle, with an attractive power equal to the sum of the powers of these different particles.

113. *Verification of the Law in the Case of the Moon.*—The moon is shown by observation to revolve about the earth in a period of 27.32 days, at a mean distance from the earth of 238,800 miles. If we take the formula for centrifugal force given in Art. 69,

$$f = \frac{4\pi^2 r}{t^2},$$

and substitute for r the moon's distance in feet, and for t its period of revolution in seconds, we shall find for the centrifugal force,

$$f = 0.0089 \text{ feet:}$$

MASS OF THE SUN.

that is to say, in one second the earth tends to give the moon a velocity towards itself of 0.0089 feet. Now the force of gravity on the earth's surface at the equator is 32.09 feet; and if the law of gravitation is assumed to be true, the force of gravity at the distance of the moon will be $\frac{32.09}{(60.267)^2}$ feet, since the distance of the moon from the earth is equal to 60.267 of the earth's radii. The value of this expression is found to be 0.0088 feet. The two results vary by only $\frac{1}{10000}$th of a foot: and it is therefore fair to conclude that the centrifugal force of the moon in its orbit is really counteracted by the earth's attraction.

In whatever way the law of gravitation is tested in connection with the observed motions of the heavenly bodies, the facts which come by observation are always found to be in close agreement with the results which the law demands; and it is safe to say that the truth of this law is as satisfactorily demonstrated as is that of the laws of refraction, of the laws of sound, or of the many other natural laws which depend upon observation and experiment for their ultimate proof.

114. *The Mass of the Sun.*—Let A denote the attraction exerted by the sun on the earth, and a that exerted by the earth on a body at its surface. Let M denote the mass of the sun, m that of the earth, r the radius of the earth, and R the radius of the earth's orbit. We have, then, by the law of gravitation,

$$\frac{A}{a} = \frac{M}{m} \times \frac{r^2}{R^2}.$$

But A must equal the earth's centrifugal force in its orbit, or $\frac{4\pi^2 R}{t^2}$, in which t is 365.256 days, reduced to seconds, and R is expressed in feet. We have also a equal to 32.09 feet. Substituting these values, we have,

$$\frac{M}{m} = \frac{4\pi^2 R^3}{32.09 t^2 r^2}.$$

Substituting the known values of the different quantities, we shall have

$$\frac{M}{m} = 327,000:$$

or the mass of the sun is equal to that of 327,000 earths.

The density of the sun, compared with that of the earth, being equal to the mass divided by the volume, is $\frac{327,000}{1,286,000}$. The density is therefore equal to about ¼th of that of the earth, and to about ⅘ths of that of water.

115. *Weight of Bodies at the Surface of the Sun.*—The weight of the same body at the surface of the earth and at that of the sun will be directly as the masses of the two spheres and inversely as the squares of their radii. We shall find that the weight of a body at the sun is about 27.7 times its weight at the earth; so that a body which exerts a pressure of 10 pounds at the earth would exert a pressure at the sun equal to that of 277 of the same pounds; and a man whose weight is 150 pounds would, if transported to the sun, be obliged to support in his own body a weight equivalent to about two of our tons.

116. *The Earth's Motion at Perihelion and Aphelion.*—We have already seen that the angular velocity of the earth in its orbit is the greatest at perihelion, when the earth is the nearest to the sun, and is the least at aphelion, when the earth is the farthest from the sun. This irregularity of motion is a consequence of the attraction exerted by the sun on the earth, as may be seen in

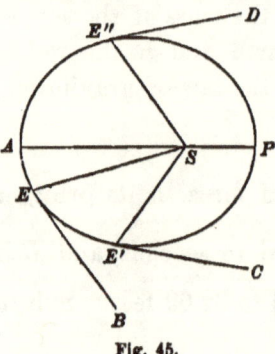

Fig. 45.

Fig. 45. In this figure, let S be the sun, P the perihelion of the earth's orbit, and A the aphelion. Let the earth move from A to P, and suppose it to be at the point E. The attraction of the sun on the earth, along the line ES, may be resolved into two forces, one of which, acting in the direction of the tangent EB, will evidently tend to increase the velocity of the earth in its orbit. At the point E''', on the contrary, where the earth is moving toward A, the effect of the sun's attraction is to diminish the earth's velocity. In general, then, the earth's velocity will increase as it moves from aphelion to perihelion, and decrease as it moves from perihelion to aphelion.

KEPLER'S LAWS.

117. In the early part of the seventeenth century, more than fifty years before the announcement by Newton of the law of universal gravitation, the astronomer Kepler, by an examination of the observations which had been made upon the motions of the planets, and which had shown that the planets revolved about the sun, discovered the following laws:—

(1.) The orbit of every planet is an ellipse, having the sun at one of its foci.

(2.) If a line, called a *radius vector*, is supposed to be drawn from the sun to any planet, the areas described by this line, as the planet revolves in its orbit, are proportional to the times.

(3.) The squares of the times of revolution of any two planets are proportional to the cubes of their mean distances from the sun.

These laws were verified by Newton in his *Principia*, in a course of mathematical reasoning, the foundation of which was his theory of universal gravitation. With regard to Kepler's first law, he showed that any two spherical bodies, mutually attracted, describe orbits about their common centre of gravity, and that these orbits are limited in form to one or another of the four conic sections,—the circle, the ellipse, the parabola, and the hyperbola. For instance, in the case of the earth and the sun, the earth does not describe an ellipse about the sun at rest, but both earth and sun revolve about their common centre of gravity. It is shown in Mechanics that the common centre of gravity of any two globes is at a point on the straight line joining their independent centres of gravity, so situated that its distances from the centres of the two globes are inversely as the masses of the globes. Hence the distance of the common centre of gravity of the earth and the sun from the centre of the sun is $\frac{92,400,000}{327,000}$ miles, or only about 280 miles; so that the sun may practically be considered to be at rest.

With regard to Kepler's second law, Newton further showed that the angular velocity with which the radius vector moves must be inversely proportional to the square of its length.

Finally, he proved also the truth of the third law, provided only that a slight correction is introduced when the mass of the planet is not so small as to be inappreciable in comparison with that of the sun. If t and t' denote the times of revolution of any two planets, m and m' their masses (the mass of the sun being unity), and d and d' their mean distances from the sun, Newton showed that we shall always have the following proportion:

$$t^2 : t'^2 = \frac{d^3}{1+m} : \frac{d'^3}{1+m'}.$$

If m and m' are so small that they may be omitted without sensible error, this proportion becomes identical with Kepler's third law.

PERTURBATIONS IN THE EARTH'S ORBIT.

118. *Precession.* — Although the absolute positions of the planes of the ecliptic and the equinoctial in space, and their relative positions to each other, remain very nearly the same from year to year, there are nevertheless certain small perturbations in these positions which are made evident to us by refined and extended observations. The principal of these perturbations is called *precession*.

The latitudes of all the fixed stars remain very nearly the same from year to year, and even from century to century: and we therefore conclude that the position of the ecliptic with reference to the celestial sphere remains very nearly unchanged. But the longitudes of the stars are all found to *increase* by an annual amount of 50″.2: and hence the line of the equinoxes must have an annual *westward* motion of the same amount. This westward motion is called the *precession of the equinoxes*. Since the ecliptic remains stationary in the heavens (or at least so nearly stationary that the latitudes of the stars only vary by half a second of arc in a year), this precession must be considered to be a motion of the equinoctial on the ecliptic.

In Fig. 46, let EQ represent the equinoctial, and LC the ecliptic. BV is the line of the equinoxes, which moves about in the plane of the ecliptic, taking in course of time the new position $B'V'$. Perhaps the clearest conception of this motion is obtained by considering P, the pole of the equinoctial, to revolve about A, the pole of the ecliptic, in the circle PG (the polar radius of which, AP, is 23° 27′), moving westward in this circle with the annual rate of 50″.2, and completing its revolution in 25,817 years.

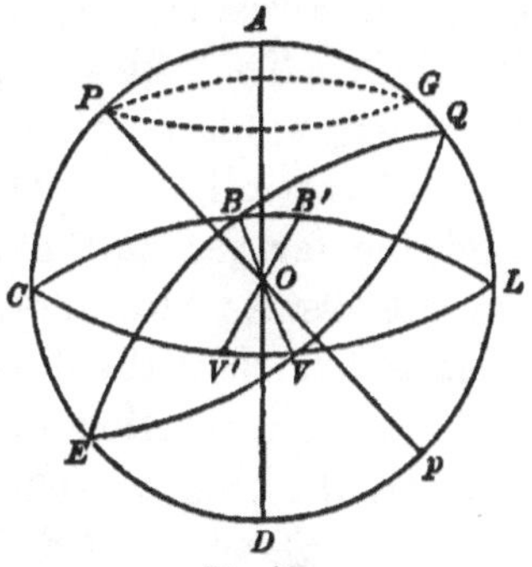

A general explanation of the cause of precession may be given by means of Fig. 47. The earth may be regarded as a sphere surrounded by a spheroidal shell, as

Fig. 46.

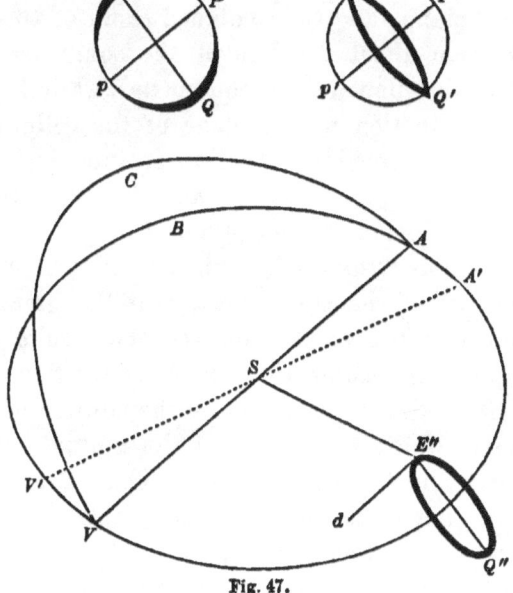

Fig. 47.

represented in the figure by $EPQp$, and the matter in this shell may be considered to form a ring about the earth in the plane of the equator, as shown in $E'P'Q'p'$. It is to the attraction of the sun and the moon on this ring, combined with the earth's rotation, that the precession of the equinoxes is due. In the figure let S be the sun, the circle ABV the ecliptic, and $E''Q''$ this equatorial ring of the earth. Let ACV be the plane of the equinoctial, meeting the plane of the ecliptic in the line of equinoxes AV. This plane is by definition determined at each instant by the position of the earth's equator. The attraction exerted by the sun upon the different particles of the ring in that half of it which is nearer the sun (the particle E'', for instance), may be resolved into two forces, one acting in the plane of the equator, and the other in a direction perpendicular to that plane, or in the direction $E''d$. The sun's attraction upon the nearer half of the ring, then, tends to draw the plane of the ring nearer to the plane of the ecliptic. On the other hand, the sun's attraction upon the farther half of the ring tends to bring about the opposite result; but since, by the law of attraction, the latter effect is less than the former, we may consider the whole result of the sun's attraction upon the ring to be a tendency in the plane of the ring to come nearer to the plane of the ecliptic. Therefore, *if the ring did not rotate*, the plane of the earth's equator would ultimately come into coincidence with the plane of the ecliptic.

But the ring does rotate, about an axis perpendicular to its own plane; and the combined result of this rotation and of the rotation about the line of the equinoxes, above described, is that the plane of the equinoctial, while it preserves constantly its inclination to the plane of the ecliptic, moves about in a westerly direction: the line of intersection of the two planes also moving about in the same direction, and thus giving rise to the precession of the equinoxes.

Similar results will evidently follow if S represents the moon instead of the sun. Owing to the greater proximity of the moon to the earth, however, the results of its attraction are more than double those of the attraction of the sun. There is still another perturbation in the position of the line of the equinoxes which is a result of the mutual attraction between the earth and the other planets. This attraction tends to draw the earth out of the plane of the ecliptic, without affecting in any way the position of the plane of the equinoctial. The result is an annual movement of the equinoxes towards the east. This perturbation is exceedingly minute, being only about $\frac{1}{8}$th of a second of arc in a year. The value $50''.2$ is the algebraic sum of all these perturbations.

119. *Results of Precession.*—One result of precession is to make the interval of time between two successive returns of the sun to the vernal equinox *less* than the time of one sidereal revolution, by the time required by the sun to pass over $50''.2$, which is 20m. 21.8s. Hence we have the *tropical* year, to which

reference has already been made in Art. 111. Another result is that the signs of the Zodiac (Art. 91) no longer coincide with the constellations after which they are named, but have retreated towards the west by about 28°, or nearly one sign: so that the constellation of Aries now lies in the sign of Taurus. Still another result is that the same star is not the pole-star in different ages. Referring to Fig. 46, the pole of the heavens, P, will have revolved about A to the position G, in the course of about 13,000 years; and a star of the first magnitude, called Vega, which is now about 51° from the pole, will at that time be less than 5° from the pole, and will be the pole-star.

120. *Nutation.*—Since precession is the result of the tendency of the sun to change the position of the plane of the equator, it is evident that there will be no precession when the sun is itself in the plane of the equator,—in other words, at the equinoxes,—and that the precession will be at its maximum when the sun is the farthest from the plane of the equator: that is to say, is at the solstices. The amount of precession due to the influence of the moon is subject to a similar variation, being the greatest when the moon's declination is the greatest. The result is that the pole of the heavens has a small oscillatory motion about its *mean* place. This motion is called *nutation*. If the effect of nutation could be separated from that of precession, the pole would be found to move in a very minute ellipse, having a major axis of 18″.5, and a minor axis of 13″.7, the period of one revolution in this ellipse being about nineteen years. Since, however, these two perturbations co-exist, the result is that the pole of the heavens revolves about the pole of the ecliptic, not in a circle, but in an undulating curve, as represented in Fig. 48: the amount of the deviation being very much exaggerated in the figure.

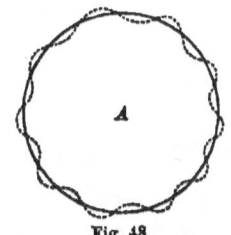

Fig. 48.

121. *Change in the Obliquity of the Ecliptic.*—It was stated in Art. 118 that the latitudes of the stars were found to vary from year to year by a very minute amount. This change in the latitudes is due to a change in the position of the plane of the ecliptic, involving a change in the obliquity

of the ecliptic. The obliquity of the ecliptic in 1878 was 23° 27′ 18″: and the annual amount of diminution to which it is now subject is 0″.46. Mathematical investigations show that after certain moderate limits have been reached this diminution will cease, and the obliquity will begin to increase. The arc through which the obliquity oscillates is about 1° 21′, and the time of one oscillation is about ten thousand years.

122. *Advance of the Line of Apsides.*—The line connecting the earth's perihelion and aphelion is called *the line of apsides.* This line revolves from west to east, with an annual rate of 11″.8; a perturbation due to the attraction exerted on the earth by the superior planets. The time in which the earth moves from perihelion to perihelion is called the *anomalistic* year (from anomaly, Art. 98). It is evidently longer than the sidereal year, and is found to contain 365d. 6h. 13m. 49.3s.

ABERRATION.

123. The *apparent* direction of a celestial body is determined by the direction of the telescope through which it is observed. In consequence of the motion of the earth, and the progressive motion of light, the telescope is carried to a new position while the light is descending through it, and therefore the apparent direction of the body will differ from its true direction.

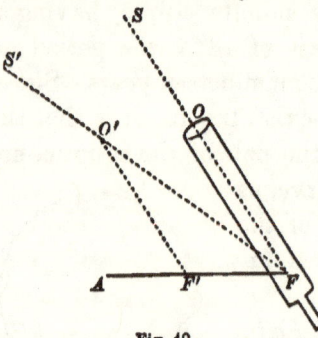

Fig. 49.

In Fig. 49, let OF be the position of the axis of a telescope at the instant when the rays of light from the star S reach the object glass O. The rays, after passing through the glass, begin to converge towards *a fixed point in space*, with which, at this instant, the intersection of the cross-wires coincides. Let the earth be moving in the direction FA. Since the transmission of light is not instantaneous, time is required for the light to pass from O to the fixed point in space, and in that time the earth will carry the axis of the telescope to some new position $O'F'$. The cross-wires will then be at F', while the rays, whose motion in space is entirely independent of any motion of the telescope, will tend to meet at the point F. In order, then, to have the image of the star coincide with the intersection of the wires, the telescope must be so moved that its axis will lie in the position $O'F$. The star will then *appear* to lie in the direction FS', while its true direction is of course FS: and the angle which these two directions make with each other, or the angle $F'O'F$, is called the *aberration*. Representing this angle by A, and the angle $O'FF'$, the angle between the apparent direction of the star and the direction in which the earth is moving, by I, we shall have,

$$\sin A : \sin I = FF' : O'F'.$$

But the ratio $FF' : O'F'$ is the ratio between the velocity of the earth and that of light: so that the sine of the aberration is equal to the ratio of the velocity of the earth to that of light, multiplied by the sine of the angle I.

124. *Diurnal Aberration.*—Aberration causes the celestial bodies to appear to be nearer than they really are to that point of the celestial sphere towards which the motion of the earth is directed at the instant of observation. As a correction, then, it is to be applied in the opposite direction. There is evidently no aberration when the motion of the earth is directly towards the star, and the greatest amount of aberration occurs when the direction of the earth's motion is at right angles to the direction of the star.

Aberration is of two kinds, corresponding to the daily and the yearly motion of the earth. The *diurnal* aberration tends to displace all bodies in the direction in which the earth is carrying the observer: that is to say, in an easterly direction. It evidently varies with the linear velocity of the observer, and is therefore the greatest at the equator and zero at the poles. Owing to the minuteness of the velocity of any point of the earth's surface about the axis in comparison with the velocity of light, the diurnal aberration is extremely small, its greatest value being less than $\frac{1}{3}$d of a second of arc.

125. *Annual Aberration.*—The displacement of a star occasioned by the motion of the earth in its orbit about the sun is called the *annual* aberration. The effect which it has on the motion of any body will depend on the relative situation of that body

to the plane of the ecliptic, as may be seen in Fig 50. In this figure S represents the sun, $ABCD$ the orbit of the earth, and K the pole of the ecliptic. Suppose a star to be at K. As the earth moves through A, in the direction indicated by the arrow, the star will be displaced from K to a; as the earth moves through B, the star will be seen at b, &c. Since the direction of this star is always at right angles to the direction in which the earth is moving, the aberration will continually be at its maximum, as shown in the previous article, and the star will describe a circle about its true place as a centre. If the star is in the plane of the ecliptic, as at s, there will be no aberration when the earth is at A or C, and the aberration will be at its maximum when the earth is at B or D. The star will therefore during the year describe the arc $b'd'$, equal in value to twice the maximum of aberration, and having the true place of the star at its middle point. If the star is situated between the pole and the plane of the ecliptic, it will describe an ellipse, the semi-major axis of which is the maximum of aberration, and the semi-minor axis of which increases with the latitude of the star.

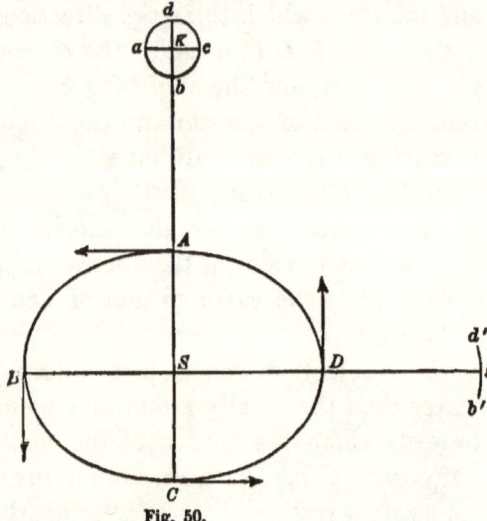

Fig. 50.

126. *Velocity of Light.*—The maximum value of aberration is the same for all bodies, and may be obtained by observing the apparent motion of a fixed star during the year. Its value has thus been obtained, and is $20''.4$. Now, since the maximum of aberration occurs when the angle I, in the formula in Art. 123, is $90°$, we shall have, denoting this maximum by A',

$$\sin A' = \frac{FF'}{O'F'}.$$

But FF' is the velocity of the earth in its orbit, or 18.4 miles a second. Hence, the velocity of light in a second is 18.4 miles multiplied by the cosecant of 20″.4, which will be found to be 185,600 miles. Experiments of a totally different character have given almost precisely the same result; and it is believed that this estimate is within a thousand miles of the true velocity.

If we divide the distance of the earth from the sun by this velocity, we find that it requires 8m. 18s. for light to pass over that distance. When we look at the sun, therefore, we see it, not as it is at the time of observation, but as it was 8m. 18s. previously; and in the same way every other celestial body appears to be in a different position from that which it really occupies at the instant we observe it. It may be well to notice here, that the time required for light to pass from any celestial body to the earth is an element in the computation of the apparent place of that body given in the Nautical Almanac. In the case of a body which changes its actual position on the celestial sphere, as a planet, for instance, allowance must also be made for what is called *planetary* aberration; since even if the earth were stationary, the apparent position of the body would be behind its true position by the amount of its motion in the time required for light to come from the body to the earth.*

127. *Aberration a Proof of the Earth's Revolution about the Sun.*—The existence of the phenomenon of aberration, as described in Art. 125, is a matter of undoubted observation: and when the close agreement of the velocity of light obtained in the preceding article with the velocity obtained by independent philosophical experiments is taken into consideration, it is fair to regard the existence of aberration as a strong direct proof of the revolution of the earth about the sun. Another proof, similar in many respects to this, will be noticed when we come to the subject of the eclipses of Jupiter's satellites.

---

* Herschel suggests (*Outlines of Astronomy*, § 335) that this might be called *the equation of light*, in order to prevent its being confounded with the real aberration of light.

CHAPTER IX.

THE MOON.

128. *The Orbit of the Moon.*—While the moon, in common with all the celestial bodies, has the apparent westward motion which is due to the rotation of the earth, it also changes its *relative* position to the other bodies, and is continually falling behind, or to the east of them, in this diurnal motion. In other words, it has an independent motion, either real or apparent, from west to east. This eastward motion is so rapid, that we only need to observe the relative situations of the moon and some conspicuous star, during a few hours on any favorable night, to notice a perceptible change in their angular distance. If the right ascension and the declination of the moon are determined from day to day, precisely as the same elements of the sun's position were determined (Art. 89), and the corresponding positions are laid down upon a celestial globe, we shall find that the moon makes a complete revolution in the heavens, about the earth as a centre, in an average period of 27d. 7h. 43m. 11.5s. We shall also find that the plane of the moon's orbit intersects the plane of the ecliptic at an angle whose mean value is 5° 8′ 44″, and in a line which, like the earth's line of equinoxes, is continually revolving towards the west: so that the apparent orbit of the moon is not a circle, but a kind of spiral. This revolution is much more rapid, however, in the case of the moon, the amount of retrogradation being about 1° 27′ in a month, and the complete revolution being effected in 18.6 years.

The movement of the moon in its orbit is represented in Fig. 51. Let E be the earth, and the circle $MANC$ the plane of the ecliptic. Let the moon be at M at any time. Then will MN be the line in which the plane of the moon's orbit intersects the plane of the ecliptic. This line is called the *line of the nodes*. That extremity of the line through which the moon passes in moving from the southern to the northern side of the ecliptic is called the *ascending node*, the other the *descending node*. Let the moon move on from M in the arc MB. When it descends to the ecliptic, it will meet it, not at the point N, but at some point N', and the line of the nodes will take the new position $N'M'$. The moon moves on to the other side of the ecliptic, passes through the arc $N'D$, and when it again returns to the ecliptic, will meet it, not at M', but at some point M'', and the line of the nodes will take the position $M''N''$. The revolution of the line of the nodes is evidently in an opposite direction to that in which the moon itself revolves, and is therefore from east to west.

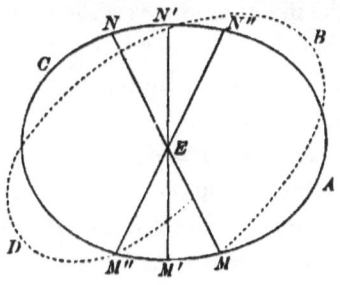

Fig. 51.

129. *Cause of the Retrogradation of the Nodes.*—This retrograde movement of the moon's nodes is similar in character to the precession of the equinoxes, and is due to the attraction which the sun exerts upon the moon. Since the plane of the moon's orbit is inclined to the plane of the ecliptic, the attraction of the sun will, in general, tend to draw the moon out of its orbit towards the ecliptic. The only exceptions to this rule will occur when the moon is at one of its nodes, and is therefore in the plane of the ecliptic, and also when the line of the moon's nodes passes through the sun, at which time the attraction of the sun is exerted along this line, and consequently in the plane of the moon's orbit. In Fig. 51 let the moon be at B, and the sun anywhere in the ecliptic except on the line of the nodes. As the moon moves on, the sun is continually drawing it down to the ecliptic, and it will hence meet the ecliptic, not at N, but at N'. The same effect will be seen at every other position of the moon in its orbit, with only the exceptions already mentioned.

130. *Change in the Obliquity of the Moon's Orbit.*—It is evident from the same figure that the angle which the arc BN' makes with the plane of the ecliptic is greater than the angle which an arc drawn through B and N would make. The obli-

quity of the plane of the moon's orbit is therefore *increased* as the moon *approaches* the node. It may be shown in the same way that the obliquity is *diminished* as the moon *recedes from* the node. The extreme limits which this angle attains are 5° 20′ 6″ and 4° 57′ 22″.

131. *Elliptical Form of the Moon's Orbit.*—The angular diameter of the moon varies at different points of its orbit, while its mean value remains the same from month to month; we therefore conclude, as we concluded in the case of the sun, that its distance from the earth is not constant, the greatest distance corresponding to the least diameter, and the least distance to the greatest diameter. If we neglect the retrogradation of the moon's nodes, and represent graphically the moon's orbit by a method identical with the method employed in representing the earth's orbit (Art. 98), we shall find the orbit to be an ellipse, with the earth at one of the foci. The eccentricity of the ellipse is 0.0549, or very nearly $\frac{1}{18}$th.

132. *Line of Apsides.*—That point in the moon's orbit where it is the nearest to the earth is called the *perigee*, and that point where it is the farthest from the earth, the *apogee*. The line connecting these two points is called the *line of apsides*. It is also the major axis of the moon's orbit. This line revolves in the plane of the moon's orbit from west to east, making a complete revolution in very nearly nine years.

The following description of the moon's orbit, and of the changes to which it is subject, is given by Herschel in his *Outlines of Astronomy*. "The best way to form a distinct conception of the moon's motion is to regard it as describing an ellipse about the earth in the focus, and at the same time to regard this ellipse itself to be in a twofold state of revolution; 1st, in its own plane, by a continual advance of its own axis in that plane; and 2dly, by a continual *tilting* motion of the plane itself, exactly similar to, but much more rapid than, that of the earth's equator."

133. *Variation in the Moon's Meridian Zenith Distance.*—From the formula in Art. 76 we have,

$$z = L - d$$

At any place, then, the latitude remaining constant, the least meridian zenith distance will occur when the moon's declination has the same name as the latitude, and is at its maximum; and the greatest will occur when the declination has the opposite name, and is also at its maximum. Since the plane of the moon's orbit is inclined, at the most, 5° 20′ to the plane of the ecliptic (Art. 130), the greatest value of the declination, either north or south, is 5° 20′ + 23° 27′. The variation in the meridian altitude will therefore be double this amount, or 57° 34′. At Annapolis, in latitude 38° 59′ N., the greatest altitude is 79° 48′, the least 22° 14′. There is an exception to this general rule in the case of those places whose latitude is less than 28° 47′: since at those places the greatest altitude occurs when the moon is in the zenith, or, as is evident from the formula, when the declination is equal to the latitude, and has the same name.

The new moon is in the same part of the heavens that the sun is in (Art. 139), and the full moon is in the opposite part. Since the sun attains its least altitude in winter and its greatest in summer, new moons will run low in winter and full moons will run high: while in summer the opposite of this will take place.

DISTANCE, SIZE, AND MASS OF THE MOON.

134. Since the sine of the moon's horizontal parallax is the

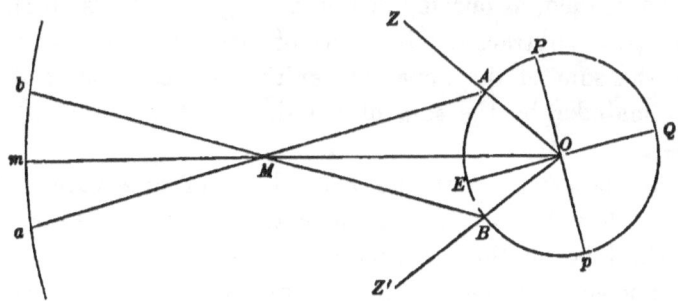

Fig. 52.

ratio of the radius of the earth to the distance of the moon from the earth, it is evident that we can determine this distance as soon as we obtain the horizontal parallax. The horizontal parallax may be found in the following manner:—In Fig. 52, let O be the centre of the earth, EQ its equator, and A and B the positions of two observers on the same meridian, whose zeniths are Z and Z'. Let M be the moon's position when crossing the meridian. The apparent zenith distance at A, corrected for refraction, is the angle ZAM, the geocentric zenith distance is ZOM, and the difference of these two angles, the angle AMO, is the parallax in altitude. In the same way OMB is the parallax at B. Represent the parallax at A by p, that at B by p', the horizontal parallax by P, the apparent meridian zenith distance at A by z, and that at B by z'.

We have, by Geometry,
$$p = z - AOM,$$
$$p' = z' - BOM,$$
and, consequently,
$$p + p' = z + z' - AOB.$$
We have also, from Art. 54,
$$p = P \sin z,$$
$$p' = P \sin z',$$
and, therefore,
$$p + p' = P (\sin z + \sin z').$$
Combining the two equations, and finding the expression for P,
$$P = \frac{z + z' - AOB}{\sin z + \sin z'}.$$
But AOB is evidently the difference of latitude of A and B. We have, then, as our method of finding the moon's horizontal parallax, to subtract the difference of latitude of the two places from the sum of the apparent zenith distances, and to divide the remainder by the sum of the sines of the two zenith distances.

It is important that the two places of observation shall differ widely in latitude. It is not, however, necessary that they shall be on the same meridian, since, either from tables of the moon's motion, or from actual observation on successive days before and after the time of observation, we can obtain the change of meridian zenith distance corresponding to any known difference of longitude, and thus reduce the two observed zenith distances to the same meridian.

135. *The Moon's Horizontal Parallax.*—By observations similar to those above described, the mean value of the moon's equatorial horizontal parallax is found to be 57' 3". The mean

distance of the moon from the earth is therefore 3962.8 miles multiplied by the cosecant of 57' 3", or 238,800 miles. The horizontal parallax varies between the limits of 61' 32" and 52' 50", and the distance between 257,900 and 221,400 miles. It must be noticed that by the mean value of the horizontal parallax given above is not meant the half sum of the two extreme values, but the value which the parallax has when the moon is at its mean distance from the earth.

136. *Magnitude of the Moon.*—The angular semi-diameter of the moon at its mean distance from the earth is found to be 15' 33".5. Its linear semi-diameter is therefore obtained by multiplying the mean distance by the sine of 15' 33".5, and is found to be 1080.8 miles, or about $\frac{3}{11}$ths of the radius of the earth. The volumes of two spheres being to each other as the cubes of their radii, the volume of the moon will be found to be about $\frac{1}{48}$th of that of the earth.

137. *Mass of the Moon.*—The mass of the moon is obtained by the following considerations. If the sun does not affect the gravity of the moon to the earth, and the mass of the moon is inappreciable in comparison with that of the earth, then the centrifugal force of the moon in its orbit ought exactly to equal the attraction of the earth on the moon. But if the moon has a sensible mass, it will, by the law of gravitation, attract the earth, and its centrifugal force must be sufficient to counterbalance the sum of the mutual attractions of the earth and the moon. Now, if we refer to what was demonstrated in Art. 113, we find that the moon's actual centrifugal force *is* greater, by $\frac{1}{88}$th, than the attraction of the earth at the distance of the moon. This would make the mass of the moon $\frac{1}{88}$th of that of the earth. But it is found that the sun diminishes sensibly the gravity of the moon to the earth. This, therefore, must also be taken into account, and the resulting value of the mass of the moon is found to be about $\frac{1}{81}$st of that of the earth.

The density of the moon, being directly as the mass, and inversely as the volume, will be $\frac{49}{81}$, or about $\frac{3}{5}$ths of the density of the earth.

138. *Augmentation of the Moon's Semi-Diameter.*—If at any time we measure the angular semi-diameter of the moon, we

shall find that it increases with the moon's altitude, being least when the moon is in the horizon and greatest when in the zenith. This increase is explained in Fig. 53. Let E be the centre of the earth, and M that of the moon. With the distance between E and M as a radius, describe the semi-circumference $AM'B$. When the moon is in the horizon of the point C, its distances from C and from E are very nearly equal. But as the moon rises, the distance CM continually decreases, while EM, the distance of the moon from the earth's centre, remains sensibly constant. When the moon is in the zenith, or at M', the distance CM' is less than EM' by the radius of the earth. Now, the angular semi-diameter of the moon will increase, as shown in Art. 98, very nearly as the distance of the moon from the observer decreases. But the earth's radius is about $\frac{1}{60}$th of the distance of the moon from the earth's centre: therefore the semi-diameter of the moon in the zenith will be greater than the semi-diameter in the horizon by $\frac{1}{60}$th of itself, or by about 15″. This increase is called the *augmentation* of the moon's semi-diameter.

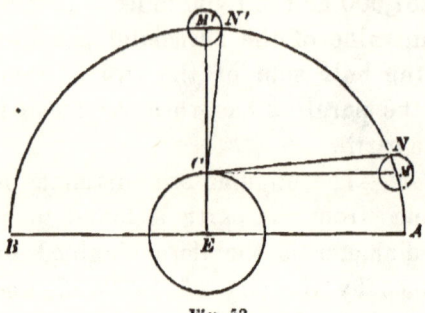

Fig. 53.

THE MOON'S PHASES.

139. Two bodies are said to be *in conjunction* when they have the same longitude. They are said to be *in opposition* when their longitudes differ by 180°; and *in quadrature* when their longitudes differ by either 90° or 270°.

The moon is an opaque body, which is rendered visible to us by the rays of light which it reflects from the sun. The *phases* of the moon are due to the different relative positions to the sun and the earth which it has while revolving about the earth.

In Fig. 54 let E be the earth, and the circle $ACFH$ the orbit of the moon. Since the inclination of the plane of the moon's orbit to the plane of the ecliptic is only a few degrees, we may neglect it in this case, and suppose the two planes to coincide. Let the sun lie in the direction *ES*. Since the distance of the

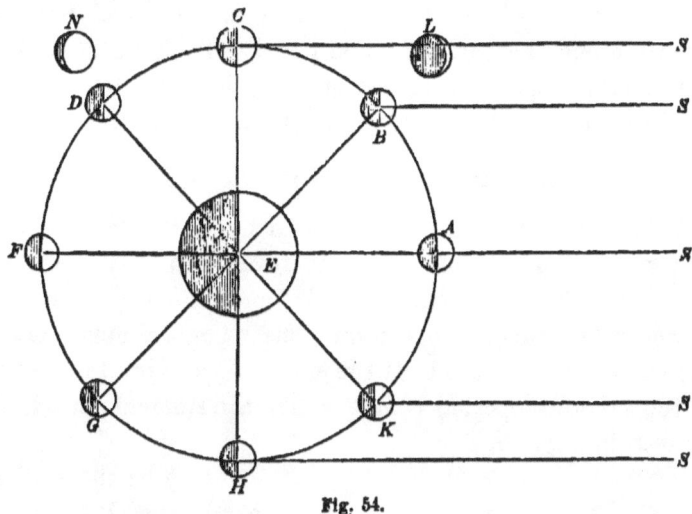

Fig. 54.

sun from the earth is about 387 times the distance of the moon from the earth, the lines *ES, HS, BS*, &c., drawn to the sun from different points of the moon's orbit, may be considered to be sensibly parallel. Let us first suppose the moon to be in conjunction with the sun at the point *A*. Here only the dark portion of the moon is turned towards the earth, and the moon is therefore invisible. This is called *new moon*. As the moon moves on towards *B*, the enlightened part begins to be visible, and when it reaches *C*, 90° in longitude from the sun, half the enlightened part is visible, and the moon is at its *first quarter*. When the moon is at *F*, in opposition to the sun, all the illuminated part is turned towards the earth, and the moon is *full*. The moon wanes after leaving *F*, passes through its *last quarter* at *H*, and finally becomes again invisible.

Between *A* and *C* the moon is *crescent*, as represented at *L*, and between *C* and *F* it is *gibbous*, as represented at *N*. The same terms are also applied to the appearance of the moon between *H* and *A* and between *F* and *H*.

140. *Phases of the Earth to the Moon.*—It is evident from Fig. 54 that the earth presents phases to the moon identical in character with those presented by the moon to the earth, although

similar phases are not presented by each body at the same time. Thus at the time of new moon the earth is *full* to the moon: and the light which it then reflects to the moon renders the unenlightened part of the moon faintly visible to the earth. As the moon moves on to its first quarter, the earth reflects less and less light to it, until finally the unenlightened portion disappears.

SIDEREAL AND SYNODICAL PERIODS.

141. The *sidereal period* of the moon is the interval of time in which it makes one complete revolution in its orbit about the earth. The *synodical period* (or *lunation*) is the interval between two successive conjunctions or oppositions. Owing to the earth's revolving about the sun, and carrying the moon with it, the synodical period is longer than the sidereal period, as may be seen in Fig. 55.

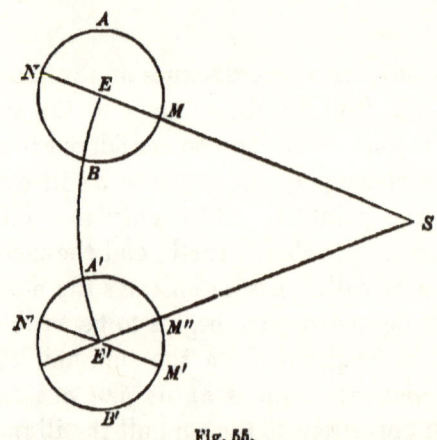

Fig. 55.

Let S be the sun, E the earth, and $MANB$ the orbit of the moon. Let the moon be at M, in conjunction with the sun. As the moon moves about E in the curve $MANB$, the earth also moves about the sun in the direction EE'. The next conjunction will therefore not occur until the moon reaches M''. Now, if through E' we draw the line $M'N'$ parallel to MN, the sidereal period of the moon is completed when the moon reaches M'. The synodical period is therefore greater than the sidereal period by the time required by the moon to pass through the angle $M'E'M''$. This angle is evidently equal to the angle ESE', which is the angular advance of the earth in its orbit in the period of one synodical revolution of the moon. In one lunar month, then, the angular advance of the moon in its orbit is greater by 360° than the angular advance of the earth in its orbit. If, therefore, we denote the moon's sidereal period in days by P, its synodical period by

S, and the earth's sidereal period, or one sidereal year, by T, we shall have,

$\dfrac{360°}{T} =$ the earth's daily angular velocity,

$\dfrac{360°}{P} =$ the moon's daily angular velocity,

$\dfrac{360°}{S} =$ the moon's daily angular gain on the earth.

Hence we shall have,

$$\dfrac{360°}{P} - \dfrac{360°}{T} = \dfrac{360°}{S};$$

$$\therefore P = \dfrac{ST}{S+T}.$$

The sidereal period of the moon is therefore obtained by multiplying the sidereal year by the moon's synodical period, and dividing the product by the sum of the sidereal year and the synodical period.

142. *Values of the Synodical and Sidereal Periods.*—The value of the synodical period is not constant, but varies from month to month. A *mean* value may, however, be obtained by dividing the interval of time between two oppositions, not consecutive, by the number of synodical revolutions in that interval. Now, the day, the hour, and even the probable minute, at which an opposition of the moon occurred in the year 720 B.C., were recorded by the Chaldæans; and by comparing this time with the results of recent observations, an extremely accurate value of the mean synodical period is obtained. It is found to be 29d. 12h. 44m. 3s. We have, then, for the value of the sidereal period, by the formula in the preceding article,

$$P = \dfrac{365.256 \times 29.53}{365.256 + 29.53} \text{ days}:$$

whence we obtain the value already given in Art. 128.

143. *Retardation of the Moon, and the Harvest Moon.*—The mean daily motion of the moon towards the east is about 13°, while that of the sun is, as we have already seen, about 1°: hence the moon is continually falling to the rear of the sun in apparent westward motion, and the interval of time between any two successive transits of the moon is greater than the similar

interval in the case of the sun. The moon, therefore, rises later, and sets later, day by day. This is called the *retardation* of the moon. Its amount varies considerably in value, but is on the average about fifty minutes.

The less the angle which the plane of the moon's orbit makes with the plane of the horizon, the less does the advance of the moon carry it with reference to the horizon, and, consequently, the less is the retardation of the moon in rising. Now, since the moon's orbit very nearly coincides with the ecliptic, the retardation in rising will in general be the least, when the ecliptic makes the least angle with the horizon. By reference to a celestial globe, it will be seen that the ecliptic makes the least angle with the horizon when the vernal equinox is in the eastern horizon. The least retardation in rising, therefore, occurs in each month when the moon is near the sign of Aries. This least retardation is especially noticeable when it occurs at the time of full moon. Now, when the moon is in Aries, *and full*, the sun must be in Libra, or near the autumnal equinox. This occurs about the 21st of September. About the time, then, of the full moon which occurs near the 21st of September, the moon will rise, for two or three nights, only about half an hour later each night. Usually this small retardation is noticed at the times of two full moons, one in September and the other in October. The first is called the *Harvest Moon*, the second the *Hunter's Moon*. All this relates to the *Northern* Hemisphere.

ROTATION. LIBRATIONS AND OTHER PERTURBATIONS.

144. *Rotation of the Moon.*—By observation of the spots upon the disc of the moon, it is found that very nearly the same surface of the moon is turned continually towards the earth. The conclusion drawn from this fact is that the moon rotates upon an axis in the same time in which it revolves about the earth, or in 27.3 days. The plane in which this rotation is performed makes an angle of about 1° 32' with the plane of the ecliptic.

If there are any inhabitants of the moon, their day will be equal in length to about twenty-nine of our days, and their night to about twenty-nine of our nights. Since the plane of the moon's equator is so nearly coincident with the plane of the ecliptic, there will hardly be any sensible change of seasons: or if there is, the lunar day will be the lunar summer, and the night the winter. To the inhabitants of one hemisphere the earth will be perpetually invisible, while to the inhabitants of the other hemisphere it will present the appearance of a body very nearly stationary in their sky, exhibiting phases similar to those which we see in the moon, with a radius nearly four times that of the moon, and a surface about thirteen times that of the moon.

145. *Librations.*—By *libration* is meant an apparent oscillatory movement of the moon, which enables us, in the course of its revolution, to see something more than an exact hemisphere.

The *libration in longitude* is due to the fact that the moon's rotation on its axis is perfectly uniform, while its motion about the earth is not. Hence the line drawn from the centre of the

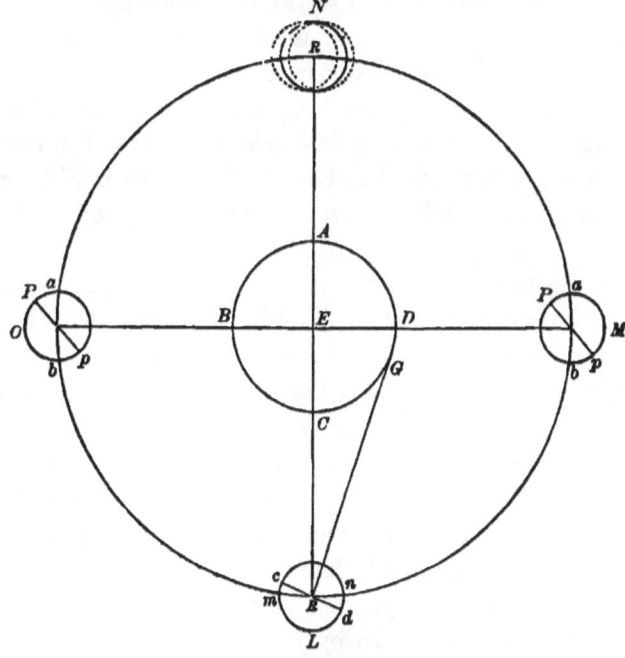

Fig. 56.

earth to that of the moon does not always intersect the surface of the moon at the same point, and we are able at times to look a few degrees, east or west, beyond the *mean* visible border. If, in Fig. 56, *ABCD* represents the earth, *E* its centre, and *R* the centre of the moon, the dotted lines at *N* denote the limits between which, as the moon revolves about the earth, the visible border may deviate from its mean position.

The *libration in latitude* is due to the fact that the axis of the moon, remaining constantly parallel to itself, is not perpendicular to the plane of the moon's orbit, but is inclined to it at an angle of about 83° 19′. We are therefore able at certain times to see about 6° 41′ beyond the north pole of the moon, and at other times the same amount beyond the south pole. Thus in Fig 56, when the moon is at *M*, we can see beyond the pole *P*, and when the moon is at *O*, beyond the pole *p*: since in each case we can see nearly that portion of the moon which lies between the earth and the circle *ab*, whose plane is perpendicular to the plane of the moon's orbit.

The *diurnal libration* is due to the difference between that hemisphere of the moon which is turned towards the centre of the earth and that which is turned towards any point on the surface. When, for instance, the moon is at *L*, an observer at *C* will see the same hemisphere which is turned towards the earth's centre, while an observer at *G* will see a different one. The hemisphere which is turned towards any observer when the moon is rising will also be different from the one which is turned towards him when the moon is setting. It is evident in the figure that the amount of this libration varies with the angle *ERG*; that is to say, with the moon's parallax.

Notwithstanding all these librations, we are able to see in all only about $\frac{57}{100}$ths of the moon's surface, according to Arago: the remainder being continually concealed from our view.

146. *Other Perturbations.*—Besides these librations, and the perturbations already mentioned (Art. 132), there are other perturbations in the moon's longitude of which only a very brief notice can here be given. The greatest of these perturbations is called *evection*, and was discovered by Ptolemy in the second century. It arises from the variation in the eccentricity of the moon's orbit, and from the fluctuations in the general advance of the line of the apsides. By it the moon's mean longitude is alternately increased and decreased by about 1° 20'. Another perturbation in the moon's motion is called *variation*. It depends solely on the angular distance of the moon from the sun, and its maximum is 37'. The *annual equation* depends on the variable distance of the earth from the sun, and amounts to 11'. The *secular acceleration* is an increase in the moon's motion which has been going on for many centuries, at the rate of about 10" a century. This perturbation is partly due to the diminution of the eccentricity of the earth's orbit; and from what has been said on that subject in Art. 98, it is evident that this inequality will at some distant day become a *secular retardation*. [See Note, page 154.]

All of these perturbations are satisfactorily explained by the investigation of what is known as *the problem of the three bodies*, in which two bodies are supposed to revolve about their common centre of gravity, according to the law of universal gravitation, and the effects of the attraction exerted by a third body upon the motions of these two bodies are made the object of mathematical examination.

THE LUNAR CYCLE.

147. If we multiply the number of days, hours, &c., in a synodical period of the moon (Art. 142) by 235, the product will be 6939d. 16h. 27m. 50s. Now, in a period of nineteen civil years there are either 6939 days, or 6940 days, according as there are four or five leap years in that period. If, then, in any year, new moon occurs on any particular day of the month, the first of January, for instance, it will occur again on the first of January (or at all events within a few hours of its end or beginning), after an interval of nineteen years; and all the new moons and the other phases will occur on very nearly the same days throughout the second period of nineteen years on which they occurred during the first period. This period is called the *Lunar Cycle*. It is also called the *Metonic Cycle*, having been originally discovered, B.C. 432, by Meton, an Athenian mathematician. The present lunar cycle began in 1862.

This cycle is used in finding Easter: Easter being the first Sunday after the full moon which occurs either upon or next after the 21st day of March.

The *golden number* of any year is the number which marks the place of that year in the cycle. It may be found for any year by adding 1 to the number of that year, and dividing the sum by 19; the remainder (or 19, if there is no remainder) is the golden number.

Four lunar cycles, or seventy-six civil years, constitute what is called the *Callippic* cycle.

GENERAL DESCRIPTION OF THE MOON.

148. When viewed through powerful telescopes, the surface of the moon is found to be made up of mountains, valleys, and plains, similar in general appearance to those that exist on the earth. As a whole, however, the surface of the moon is much more uneven than that of the earth. The heights of over 1000 lunar mountains have been measured, and some of them have been found to exceed 20,000 feet. Many of these mountains bear the appearance of having been at one time volcanoes, far surpassing in size and activity those on the earth. The common belief among astronomers seems to be that these lunar volcanoes are now extinct. Messrs. Beer and Mädler, two Prussian astronomers who have made the moon their special study, have detected no signs of activity in any of the volcanoes which they have examined. A few years ago, certain phenomena were noticed which seemed to show that one at least of these volcanoes, named Linné, is not extinct: but later observations do not confirm this suspicion.

There are no signs of the existence of water on the moon. Certain large dark patches are seen, which were formerly considered to be oceans, gulfs, &c., and were so named; but increased telescopic power shows that they are dry plains.

It seems to be still an open question whether or not the moon has an atmosphere. If there is an atmosphere, it must be of an extremely minute height and density; for we see no clouds and no twilight, and there is nothing in the phenomena of the occultations of stars by the moon which shows the existence of even the rarest atmosphere. Some observers, however, and among them Messrs. Beer and Mädler, believe that they have detected signs of the existence of a very slight atmosphere.

UNIV. O
CALIFORN

PLATE II

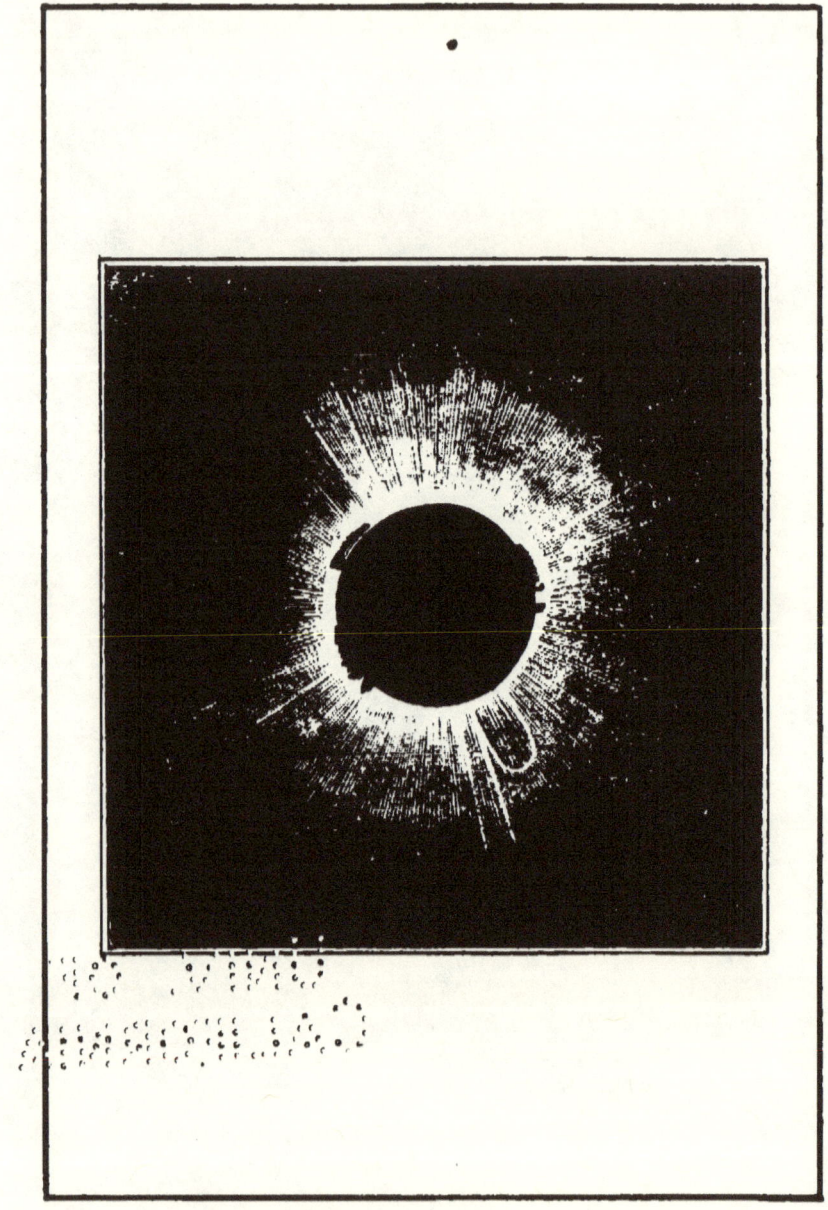

TOTAL ECLIPSE OF THE SUN, OF JULY 18, 1860,
showing the Corona and the Red Flames: as observed by Dr. Feilitzsch, at Castellon de la Plana.

CHAPTER X.

LUNAR AND SOLAR ECLIPSES. OCCULTATIONS.

149. *Eclipses.*—The obscuration, either partial or total, of the light of one celestial body by another is in astronomy termed an *eclipse.* When the earth comes between the sun and the moon, the light of the sun is shut off from the moon, and we have a *lunar eclipse.* A lunar eclipse can occur only at the time when the moon is in opposition to the sun, that is to say, at the time of full moon. When the moon comes between the earth and the sun, the light of the sun is shut off from the earth, and we have a *solar eclipse.* A solar eclipse can occur only at the time of new moon. An eclipse of a star or a planet by the moon is called an *occultation.*

If the orbit of the moon lay in the plane of the ecliptic, a lunar and a solar eclipse would occur in every month. Owing, however, to the inclination of the plane of the moon's orbit to the plane of the ecliptic, the latitude of the moon is usually too great to allow either kind of eclipse to take place; and it is only in special cases, when the moon is in or near the plane of the ecliptic at the time of conjunction or opposition, that an eclipse of the sun or the moon is possible.

LUNAR ECLIPSE.

150. In Fig. 57 let *S* be the centre of the sun, and *E* that of

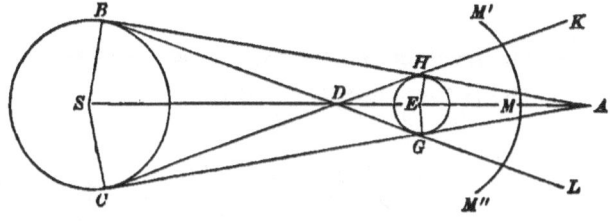

Fig. 57.

136   LUNAR ECLIPSE.

the earth. Draw the lines BH, and CG, tangent to the two spheres. These lines will meet at some point A, and $AHEG$ will be a section of the shadow cast by the earth.

The whole shadow is of a conical shape, the vertex of the cone being at A; and a lunar eclipse will occur whenever the moon is within this shadow. Draw the tangent lines BG and CH. KDL is a section of a second cone whose vertex is at D. The earth's shadow is called the *umbra*, and that portion of the second cone which lies outside of the umbra is called the *penumbra*. Thus KHA and AGL are sections of the penumbra. It must be noticed, in regard to the construction of this figure, that since only one tangent can be drawn to the circumference of a circle at any one point, the lines BG and CH do not touch the two circumferences at precisely the same points at which BH and CG touch; and that, furthermore, in all these figures the relative size of the sun should be immensely greater than it is.

Now let $M'MM''$ represent a portion of the moon's orbit at the time of a lunar eclipse. As soon as the moon passes within the line DK, some of the rays of the sun will be cut off from it by the earth, and its brightness will begin to decrease. The whole disc, however, will still be visible. As soon as the moon begins to pass within the line HA, the disc will begin to disappear, and when the whole disc has passed within the cone, the eclipse will be total.

151. *Different Kinds of Eclipses.*—When the moon's orbit is so situated that only a part of the moon enters the umbra, we have a *partial eclipse*. When the moon does not enter the umbra, but merely touches it, we have an *appulse*. When the centre of the moon coincides with the line which connects the centre of the earth and that of the sun, the eclipse is *central*. A central eclipse occurs very rarely, if indeed it occurs at all.

152. *The Semi-Angle of the Umbral Cone.*—The semi-angle of the umbral cone is the angle EAG, Fig. 58. Now we have, by Geometry,

$$SEC = ECG + EAG.$$

But SEC is the sun's angular semi-diameter, and ECG is its horizontal parallax. Putting S for SEC, and P for ECG, we have,

$$EAG = S - P.$$

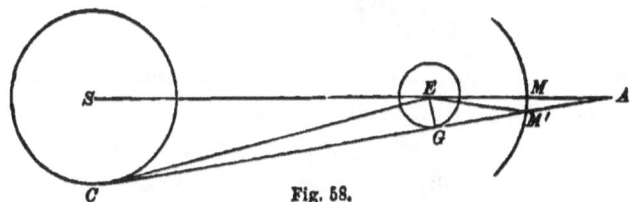

Fig. 58.

153. *The Angular Semi-Diameter of the Shadow at the Distance of the Moon.*—The angular semi-diameter of the shadow at the distance of the moon is the angle MEM'. We have, by Geometry,

$$EM'G = MEM' + EAG.$$

Now $EM'G$ is the moon's horizontal parallax, which we will represent by P', and the value of EAG has been obtained in the preceding article. We therefore have,

$$MEM' = P' + P - S.$$

Observation shows that the earth's atmosphere increases the apparent breadth of the shadow by about its one-fiftieth part: hence in practice the angular semi-diameter of the shadow is taken equal to $\frac{51}{50}(P' + P - S)$. If we substitute in this expression the least values of P' and P, and the greatest value of S, from the table given in Art. 155, we shall find that the least value of the angular semi-diameter of the shadow is about 37′ 25″: so that the entire breadth of the shadow is always more than double the greatest diameter of the moon.

154. *Length of the Earth's Shadow.*—The length of the shadow, or the line EA, can be computed from the right-angled triangle EAG, in which we have,

$$EA = EG \text{ cosec } (S - P).$$

The mean value of this length is 858,000 miles, or more than three times the distance of the moon from the earth.

155. *Lunar Ecliptic Limits.*—We see from Fig. 58 that a lunar eclipse can occur only when the moon's geocentric latitude at the time of opposition, (or at full moon,) is less than the sum of the angular semi-diameter of the shadow and the semi-diameter of the moon. If we represent the moon's semi-diameter by S, the expression for this sum is

$$\frac{51}{50}(P+P'-S)+S'.$$

If the moon's geocentric latitude at the time of opposition is greater than the greatest value which this expression can attain, no eclipse can possibly occur: if it is less than the least value of the expression, an eclipse is inevitable. These two values of this expression are called the *lunar ecliptic limits*. Now, we have by observation the following values of P, P', &c.:

	MAXIMA.	MINIMA.
P'	61′ 32″	52′ 50″
P	9	9
S'	16 46	14 24
S	16 18	15 45

In order to find the greatest value of the expression, we substitute in it the greatest values of P, P' and S', and the least value of S. The result is 1° 3′ 37″: and no eclipse will occur when the moon's latitude exceeds this limit. The least value of the expression is 51′ 49″: and when the moon's latitude at opposition is less than this, an eclipse cannot fail to occur. There are some considerations, however, which have not been taken into account, which may increase each of these limits by about 16″.

When the moon's latitude at opposition is within these limits, an eclipse is possible, but not necessarily certain. In order to determine whether in such case it will or will not occur, the actual values which P, P', S and S' will have at that time must be substituted in the expression, and the result compared with the corresponding latitude of the moon.

156. Since a lunar eclipse is caused by the moon's entering the earth's shadow, it will be seen at the same instant of time by every observer who has the moon above his horizon: and the character of the eclipse, whether total or partial, will be every-

where the same. As the moon's motion towards the east is more rapid than that of the earth (and consequently of the shadow), the eclipse will begin at the *eastern* limb of the moon. A total eclipse of the moon may last for nearly two hours. Even when totally eclipsed, however, the moon does not, in general, disappear from view, but shines with a dull reddish light. This phenomenon is caused by the earth's atmosphere, which refracts the rays of light from the sun which enter it near the points G and H, Fig. 57, and turns them into the cone. The rays which pass still nearer to these points are probably absorbed by the atmosphere, thus giving rise to the observed increase of the shadow mentioned in Art. 153.

SOLAR ECLIPSE.

157. In Fig. 59, let S represent the sun, E the earth, and M the moon, at the time of a solar eclipse. HAK will be a section of the moon's umbra, and GHA and AKD, sections of its penumbra.

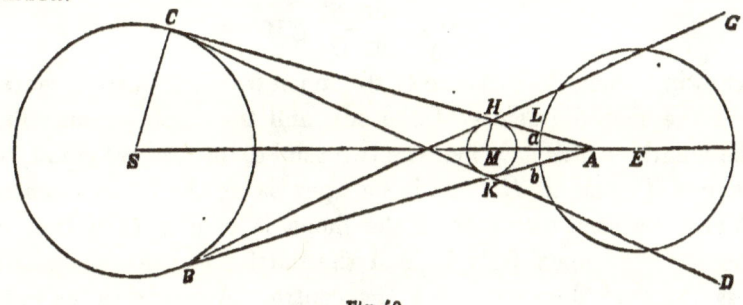

Fig. 59.

To an observer situated within the umbra, at any point of the arc ab, the eclipse will be total; while to one situated within the penumbra, as at L, for instance, the eclipse will be partial. Beyond the penumbra no eclipse whatever will be seen. Hence the geographical position of the observer determines the character of the eclipse: a condition different from that in the case of a lunar eclipse, which we have seen is the same to all observers.

158. *Length of the Moon's Shadow.*—It is evident that, to an observer at the apex of the shadow A, the angular semi-diameters of the sun and the moon would be equal. Now, the mean angular semi-diameters of these two bodies as seen from the earth's centre

are nearly equal; hence the mean position of the apex does not fall very far from the earth's centre E. An approximate value of the length of the shadow may be thus obtained. We have in Fig. 59,

$$\sin HAM = \frac{HM}{AM} = \frac{CS}{AS}.$$

But we have just now seen that HAM is the sun's angular semi-diameter, as seen from A; and as AE is small compared with AS, we may consider the angle HAM to be the sun's geocentric semi-diameter. Denoting this by S, we have,

$$\sin S = \frac{HM}{AM}.$$

Now, if S' represents the moon's *geocentric* semi-diameter, we have,

$$\sin S' = \frac{HM}{EM}.$$

Combining these two equations, and finding the value of AM, we have,

$$AM = \frac{\sin S'}{\sin S} EM.$$

Knowing, then, the distance of the moon from the earth's centre, and the semi-diameters of the sun and the moon, we may find the length of AM. When the two semi-diameters are equal, we have AM equal to EM, and the apex is at the earth's centre. When the semi-diameter of the moon is greater than that of the sun, the apex falls beyond the earth's centre: when it is less, the apex does not reach the centre. Appropriate calculations will show that when both sun and moon are at their mean distances from us, the apex falls short of the earth's surface: and that when the moon is at its least distance from the earth, and its shadow is the longest, the apex falls about 14,000 miles beyond the earth's centre.

159. *Different Kinds of Eclipses.*—When the shadow falls beyond the earth's surface, the eclipse is total, as we have already seen, within the umbra, and partial within the penumbra. When the apex just touches the earth, the eclipse is total only at the point where it touches. When the apex falls short of the surface, there will be no total eclipse; but at the point in which the axis of the cone, prolonged, meets the earth, the

observer will see what is called an *annular* eclipse, the moon being projected upon the disc of the sun, but not covering it.

160. *Solar Ecliptic Limits.*—In Fig. 60 let S represent the sun, E the earth, and M the moon. No eclipse of the sun can occur unless some part of the moon passes within the lines BC

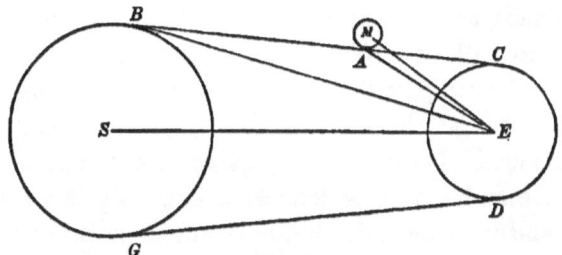

Fig. 60.

and GD, drawn tangent to the sun and the earth: that is, unless the moon's geocentric latitude is less than the angle MES. Now we have,

$$MES = MEA + AEB + BES,$$

and, also,

$$AEB = CAE - CBE.$$

BES is the sun's semi-diameter, MEA that of the moon, CAE the moon's horizontal parallax, and CBE the sun's: hence, using the notation already employed in Art. 155, we have,

$$MES = S + S' + P' - P.$$

The greatest value of this expression is found by employing the greatest values of S, S', and P', and the least value of P, as given in Art. 155, and is 1° 34′ 27″: and there will be no eclipse if the moon's latitude at conjunction is greater than this amount. The least value is 1° 22′ 50″; and if the latitude at conjunction is less than this an eclipse is inevitable. These two values, which, owing to certain considerations omitted in this discussion, should both be increased by about 25″, are called the *solar ecliptic limits*. In order to determine whether an eclipse will occur when the moon's latitude at conjunction falls within these limits, we must substitute in the expression the values which the different quantities will really have at that time, and compare the result with the corresponding latitude of the moon.

161. *General Phenomena.*—Since the moon moves towards

the east more rapidly than the sun, a solar eclipse will begin at the *western* side of the sun. For the same reason the moon's shadow will cross the earth from west to east, and the eclipse will begin earlier at the western portions of the earth's surface than at the eastern. The moon's penumbra is tangent to the earth's surface at the beginning and the end of the eclipse, so that the sun will be rising at that place where the eclipse is first seen, and setting at the place where it is last seen. A solar eclipse may last at the equator about $4\frac{1}{2}$ hours, and in these latitudes about $3\frac{1}{2}$ hours. That portion of the eclipse, however, in which the sun is wholly concealed can only last about eight minutes: and in these latitudes, only about six minutes.

The darkness during a total eclipse, though subject to some variation, is scarcely so intense as might be expected. The sky often assumes a dusky, livid color, and terrestrial objects are similarly affected. The brighter planets and some of the stars of the first magnitude generally become visible; and sometimes stars of the second magnitude are seen. The corona and the rose-colored protuberances described in Art. 102 also make their appearance. When the sun's disc has been reduced to a narrow crescent, it sometimes appears as a succession of bright points, separated by dark spaces. This phenomenon bears the name of *Baily's beads*. The dark spaces are supposed to be the lunar mountains, projected upon the sun's disc, and allowing the disc to show between them.

Occasionally the moon's disc is faintly seen, shining with a dusky light. This is caused by the rays of the sun, reflected back to the moon by that portion of the earth's surface which is still illuminated by the sun: just as at the time of new moon its entire disc is rendered visible.

CYCLE AND NUMBER OF ECLIPSES.

162. *Cycle of Eclipses.*—In order that either a solar or a lunar eclipse shall occur, it is necessary, as we have seen, that the moon shall be near the ecliptic (in other words, near the line of nodes of its orbit), at either conjunction or opposition. It is evident that when the moon is near the line of nodes at such a time, the sun also must be near the same line. The

occurrence of eclipses, then, depends on the relative situations of the sun, the moon, and the moon's nodes, and is only possible when they are all in, or nearly in, the same straight line. We have already seen (Art. 128) that the line of nodes is continually revolving to the west, completing a revolution in about 18.6 years. The sun, then, in its apparent path in the ecliptic will move from one of the moon's nodes to the same node again in less than a year. This interval of time may be called the *synodical period of the node*, and is found to be 346.62 days.

Now, we have,
$$19 \times 346.62d. = 6585.8d:$$
and, the lunar month being 29.53 days, we have also,
$$223 \times 29.53d. = 6585.2d.$$

If, then, the moon is full and at its node on any day, it will again be full, and at the same node, or very nearly at it, after an interval of 6585 days: and the eclipses which have occurred in that interval will occur again in very nearly the same order. This period of 6585 days, or 18 years and 10 days, is called the *cycle of eclipses*. It was known to the Chaldæan astronomers under the name of *Saros*. Care must be taken not to confound this cycle with the lunar cycle described in Art. 147.

163. *Number of Eclipses.*—Since the limit of the moon's latitude is greater in the case of a solar eclipse than in the case of a lunar eclipse, there are more solar eclipses than lunar eclipses. Usually 70 eclipses occur in a cycle, of which 41 are solar and 29 are lunar. Since we know that a solar eclipse is inevitable when the moon is so near the line of nodes at conjunction that its latitude is less than $1° 23' 15''$, we can compute the corresponding angular distance of the sun at the same time from this line; and having computed this, we may also determine the length of time required by the sun in passing through double this angle, or, in other words, the time required in passing from one of these limits to the corresponding limit on the other side of the same node. If we do this, we shall find that the sun cannot pass either node of the moon's orbit without being eclipsed: and therefore there must be at least two solar eclipses in a year. The greatest number that can occur is five. The greatest number of lunar eclipses in the year is three, and

144 OCCULTATIONS.

there may be none at all. The greatest number of both kinds of eclipses in a year is seven; the usual number is four.

Although the annual number of solar eclipses throughout the whole earth is the greater, yet at any one place more lunar eclipses are visible than solar. The reason of this is that a lunar eclipse, when it does occur, is visible over an entire hemisphere, while the area within which a solar eclipse is visible is very much more limited.

OCCULTATIONS.

164. An occultation of a planet or a star will occur whenever the planet or star is so situated in latitude as to allow the moon to come in between it and the earth. In order to determine the limit of a planet's latitude within which an occultation of the planet is possible, let us refer to Fig. 61. In this figure, E is the

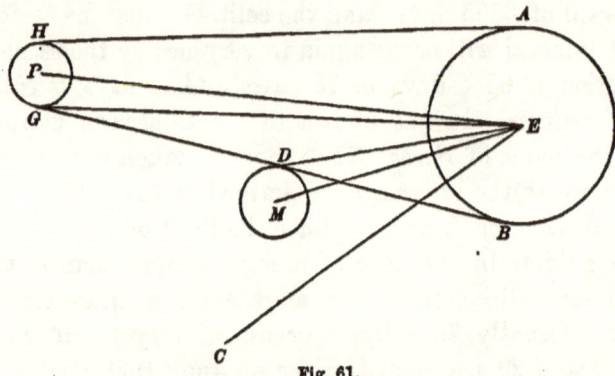

Fig. 61.

centre of the earth, P that of a planet, and M that of the moon. An occultation will occur when the moon comes between the tangent lines GB and AH. Let EC be the plane of the ecliptic. PEC is then the geocentric latitude of the planet, and MEC that of the moon.

We have,
$$PEC = PEG + GED + DEM + MEC,$$
and also,
$$GED = EDB - EGD.$$

Now, PEG is the planet's semi-diameter, EGD its horizontal parallax, DEM the moon's semi-diameter, EDB its horizontal parallax, and MEC, as above stated, its latitude. The value of *PEC*, therefore, can very readily be obtained. If P, instead of representing a planet, represents a star, the distance PE becomes so great that AH and BG are sensibly parallel, and the star's parallax and semi-diameter reduce to zero. In this case the greatest value of *PEC*, within which an occultation can occur, will be the sum of 5° 20′ 6″, 61′ 32″, and 16′ 45″, which is 6° 38′ 23″.

Since the moon moves from west to east, the occultation always takes place at its eastern limb. From new moon to full, the dark portion of the moon is to the east, as may be seen in Fig. 54, and from full moon to new, the bright limb is to the east. When an occultation occurs at the dark edge, particularly if the moon is so far on towards its first quarter that the dark portion is invisible, the disappearance is extremely striking, as the occulted body appears to be extinguished without any visible interference.

As already stated in Art. 83, a *solar* eclipse, or an occultation of a planet or star, although not visible at different places at the same absolute instant of time, may still be made the means of very accurately determining the longitude of a place, or the difference of longitude of two places. For instance, in the case of a solar eclipse, we may deduce, from the local times of the beginning and the end of the eclipse, as observed at any place, the time of true conjunction of the sun and the moon: the time of conjunction, that is to say, as seen from the centre of the earth. If, then, we compare the local time of true conjunction with the Greenwich time of true conjunction, it amounts to comparing the local and the Greenwich time corresponding to the same absolute instant: and the difference of these two times will evidently be the longitude of the place of observation from Greenwich.

CHAPTER XI.

THE TIDES.

165. The surface of the ocean rises and falls twice in the course of a lunar day, the length of which is, as we have already seen (Art. 143), about 24h. 50m. of mean solar time. The rise of the water is called *flood tide*, and the fall *ebb tide*. When the water is at its greatest height it is said to be *high water*, and when at its least height, *low water*.

166. *Cause of the Tides.*—The tides are due to the *inequality* of the attractions exerted by the moon upon the earth and the waters of the ocean, and to a similar but smaller inequality in the attractions exerted by the sun.

In order to examine the phenomena of the tides, we will consider the earth to be a solid globe, surrounded by a shell of water of uniform depth. The centrifugal force induced by the rotation of the earth would tend to give a spheroidal form to this shell of water; but as we wish simply to examine the effects of the attractions exerted by the moon and the sun, we will disregard the rotation of the earth, and consider it to be at rest.

In Fig. 62, then, let $ABCD$ represent the earth, and the dotted

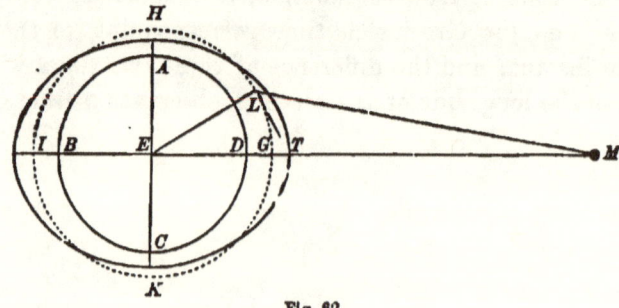

Fig. 62.

line $GHIK$ the surrounding shell of water. Let M be the moon. The attraction of the moon on the solid mass of the earth is the same that it would be if the whole mass were concentrated at the point E. Now since, by the law of gravitation, the attraction of the moon on any two particles is inversely as the square of the distances of the two particles from the moon, the attraction exerted upon the particle of water at G will be greater than that exerted upon the general mass of the earth, supposed to be concentrated at E. The particle G will therefore *tend* to recede from the earth: that is to say, its gravity towards the earth's centre will be diminished, although, as is plain, it will not move. The same result will follow at the opposite point I. The moon will exert a greater attractive power upon the mass of the earth than upon a particle at I, and will tend to draw the earth away from the particle: so that the gravity of the particle at I towards the earth's centre will also be diminished. Since the ratio of the distances ME and MG is very nearly equal to the ratio of the distances MI and ME, the amount of the decrease of gravity at G and at I will be nearly the same.

Let us next examine the effect of the moon's attraction at some point L, not situated vertically under the moon. The attraction of the moon at this point is less than that at the point G, since the distance ML is greater than the distance MG; and since the attraction exerted on the mass of the earth is, of course, the same for both points, the *difference* of the attractions exerted on the earth and the water is less at the point L than at the point G. At the point L, however, this inequality of attraction is not wholly counteracted by gravity: for if the force with which the moon tends to draw a particle at L along the line ML be resolved into two forces, one in the direction of the radius EL, and the other in the direction of the tangent LT, the latter force will cause the particle to move towards the point G. The same result will follow at any other point of the arc HGK: so that all the water in that arc tends to flow towards the point G, and to produce high water there.

In the same way it may be shown that the water in the arc HIK tends to flow towards the point I, and to produce high water at that point.

The result, then, of the attraction of the moon, exerted under the suppositions which we made at the outset, is to give to the shell of water a spheroidal form, as shown in the figure, the major axis of the spheroid being directed towards the moon. Suitable investigation shows that the difference of the major and the minor semi-axis of this spheroid is about fifty-eight inches.

167. *Daily Inequality of the Tides.*—The rotation of the earth, and the inclination of the plane of the moon's orbit to the plane of the equator, produce in general an inequality in the two daily tides at any place. In order to show this, we will suppose that the spheroidal form of the water is assumed instantaneously in each new position of the earth as it rotates. In Fig. 63, let E be the centre of the earth, surrounded, as in Fig. 62, by a spheroidal shell of water, the transverse axis of the spheroid lying in the direction of the moon M. Let Pp be the axis of rotation of the earth, and CD the equator. The angle MED is the moon's declination.

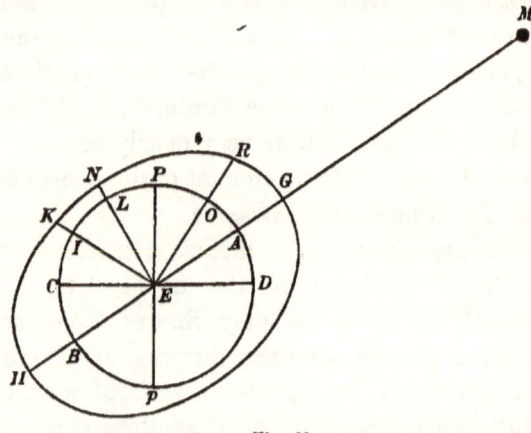

Fig. 63.

The water is at its greatest height, as before, at the points A and B, and the height at other points diminishes as the angular distance of those points from the line GH increases. Let I be a place having the same latitude that A has, but situated 180° from it in longitude. The height of the tide at I is represented by IK. In a little more than twelve hours the rotation of the earth will have caused I and A to change places with reference to the moon. I will then be where A is in the figure, and will have a tide with the height AG, while A will be where I is now, and will have a tide with the height IK. It is not necessary to prove that IK is less than AG. We see, then, that at both A and I the two daily tides are unequal, the greater of the two occurring at each place at the time of the moon's upper culmination at that place, or being, at all events, the one which occurs next after that culmination. The same daily inequality of tides may be shown to exist at any other points on the earth's surface, as, for instance, at L and O. At the equator, however, and also at the poles, the two daily tides are sensibly equal, as may readily be seen from the figure.

168. *General Laws.*—As far as the influence of the moon is concerned in causing tides, the following general laws may be deduced from what has been shown in the preceding articles.

(1.) When the moon is in the plane of the celestial equator, or, in Fig. 63, when EM coincides with ED, the tides are greatest at the equator, and diminish at other places as the latitude increases; and the two daily tides at any place are sensibly equal.

(2.) When the moon is not in the plane of the celestial equator, the two daily tides at any place except the poles and the equator are unequal. The greatest tides, and the greatest inequality of tides, occur at those places whose latitude is numerically equal to the moon's declination. If the place is on the same side of the equator as the moon, the greater of the two daily tides occurs at or next after the upper culmination of the moon; if the place is on the opposite side of the equator, the greater tide occurs at or next after the lower culmination of the moon.

(3.) Owing to the retardation of the moon (Art. 143), there is a similar retardation in the occurrence of high water. The length of the lunar day being on the average 24h. 50m., the average interval of time between two successive tides is 12h. 25m.

169. *Influence of the Sun in Causing Tides.*—All that has been said in the preceding articles with regard to the influence of the moon in creating tides is equally true with regard to the influence of the sun. The mass of the sun being so immense in comparison with that of the moon, it might be supposed that the influence of the sun over the tides would be greater than that of the moon, even although its distance from the earth is much greater than that of the moon. But such is not the case in fact. The height of the tide produced by either body is not so much due to the absolute attraction which that body exerts, as to the *relative* attractions which it exerts on the solid mass of the earth and on the water: and the moon is so much nearer to the earth than the sun, that the *difference* of its attractions on

the earth and the water is greater than the corresponding difference in the case of the sun. It is computed that the effect of the moon in creating tides is about $2\frac{1}{3}$ times that of the sun.

170. *Combined Effects of both Sun and Moon.*—Since each body, independently of the other, tends to raise the surface of the water at certain points, and to depress it at certain other points, the tides will evidently be higher when both bodies tend to raise the surface of the water at the same time, than when one tends to raise and the other to depress it. At new and full moon the two bodies act together, while at the first and the last quarter they act in opposition to each other. The tides at the former periods will therefore be the greater, and are called *spring* tides; and the tides at the latter periods are called *neap* tides. The ratio of the spring to the neap tide is that of $(2\frac{1}{3} + 1)$ to $(2\frac{1}{3} - 1)$, or of 5 to 2.

The height of the tide is also affected by the change in the distance of the attracting body. For instance, when the moon is in perigee, the tides tend to run higher than when the moon is in apogee; and when the moon is in perigee, and also either new or full, unusually high tides will occur.

171. *Priming and Lagging of the Tides.*—Each of these bodies may be supposed to raise a tidal wave of its own, and the actual high water at any place may be considered to be the result of the combination of the two waves. When the moon is in its first or its third quarter, the solar wave is to the west of the lunar one, and the actual high water will be to the west of the place at which it would have been had the moon acted alone. There is therefore at these times an acceleration of the time of high water, which is called the *priming* of the tides. In the second and the fourth quarter, the solar wave is to the east of the lunar one, and a retardation of the time of high water occurs, which is called the *lagging* of the tides.

172. Although the theory of the tides, on the supposition that the earth is wholly covered with water. admits of easy explanation, the actual phenomena which they present are very much more complicated, and must be obtained principally from observation. The lunar wave mentioned in the preceding article being greater than the solar wave, we may consider the two to-

gether to constitute one great *tidal wave*, which at every moment tends to accompany the moon in its apparent diurnal path towards the west, raising the waters at successive meridians, but giving them little or no *progressive* motion. This tidal wave would naturally move westward with an angular velocity equal to that of the moon, so that at the equator its motion would be about 1000 miles an hour; but the obstructions offered by the continents, the irregularity of their outlines, the uneven surface of the ocean bed, and the action of winds and currents and friction, all combine not only to diminish the velocity of the tidal wave, but also to make it extremely variable.

173. *Establishment of a Port.*—The interval of time between the moon's transit over any meridian and high water at that meridian varies at different places, and varies also on different days at the same place. This interval of time is called the *luni-tidal* interval. The mean of the values of this interval on the days of new and full moon is called the *common establishment* of a port. The mean of all the luni-tidal intervals in the course of the month is called the *corrected* establishment. These establishments are obtained by observation, and are given in Bowditch's Navigator, and also in other works. Thus the establishment of Annapolis is 4h. 38m., and that of Boston 11h. 12m. The time of transit of the moon over any meridian on any day can be obtained from the Nautical Almanac: and the sum of this time of transit and of the establishment of any port will be the approximate time of that flood tide which occurs next after the transit. Suppose, for instance, the time of high water at Annapolis, on January 11, 1867, is desired. We have from the Almanac the time of the moon's transit, 4h. 33m.; adding to this the establishment, 4h. 38m., the sum is 9h. 11m. This is, in this case, the time of the evening tide. The time of the morning tide may be obtained by subtracting 12h. 25m. from this time, which will give us 8h. 46m. A.M. A more accurate result in this last case might be obtained by taking from the Almanac the time of the preceding lower transit, and adding the establishment to it; but practically this would be a needless refinement, for the two results would only vary by about two minutes.

The time of transit which the Almanac gives is in astronomical time: hence the resulting time of high water will also be in astronomical time, and it will frequently happen that the time which we find, when turned into civil time, will fall on the civil day subsequent to the day for which the time is desired. Take, for instance, January 23, 1867, at Annapolis. The time of transit is January 23, 15h. 34m.: hence the time of high water is January 23, 20h. 12m., or, in civil time, January 24, 8h. 12m. A.M. If, then, we wish the time of high water on the morning of the civil day January 23d, we must take from the Almanac the time of transit for the astronomical day of January 22d.

There are other tables given in Bowditch's Navigator, and in the United States Coast Survey Reports, by which the time of high water can be obtained with greater accuracy than by the method given above.

174. *Cotidal Lines.*—If the tidal wave were everywhere uniform in its progress, it would come to all points on the same meridian at the same time. But, owing to irregularities induced by local causes, such is not the case, and places on different meridians often have high water at the same instant of time. Charts are therefore published on which are drawn lines connecting places where high tides occur at the same instant: and these lines are called *cotidal* lines. These lines are usually accompanied by numerals, which indicate the hours of Greenwich time at which high tides occur on the days of new and full moon along the different lines.

175. *Height of Tides.*—At small islands in mid-ocean the height of the tides is not great, being sometimes less than one foot. When the tidal wave approaches a continent, and the water begins to shoal, the velocity of the wave is diminished, and the height of the tide is increased. When the wave enters bays opening in the direction in which the wave is moving, the height of the tide is still further increased.

The eastern coast of the United States may be considered to constitute three great bays: the first included between Cape Sable, in Nova Scotia, and Nantucket, the second between Nantucket and Cape Hatteras, and the third between Cape Hatteras and Cape Sable, in Florida. In each of these bays the tides, in general, increase in height from the entrance of the bay to its head.

For instance, in the most southern of these bays, the tides at Cape Sable and Cape Hatteras are not more than two feet in height; while at Port Royal, at the head of this bay, they are about seven feet. The same thing is noticed, in general, in smaller bays and sounds. For instance, in Long Island Sound, the height of the tide is two feet at the eastern extremity, and more than seven feet at the western. This increase of height is particularly noticeable in the Bay of Fundy, in which the height is eighteen feet at the entrance, and fifty and sometimes seventy feet at the head.

There are in some cases, however, special causes which create exceptions to this general rule of increase of the tides between the entrance and the head of a bay. In Chesapeake Bay, for instance, which is wider at some places than it is at the entrance, and which lies about north and south, the tides in general diminish in height as we ascend the bay.

176. *Tides in Rivers.*—The same general principle holds good in the tides of rivers. When the channel contracts or shoals rapidly, the height of the tide increases: when it widens or deepens, the height decreases. In a long river, then, the tides may alternately increase and decrease. For instance, at Tivoli, on the Hudson, between West Point and Albany, the tide is higher than it is at either of those two places.

177. *Different Directions of the Tidal Wave.*—The tidal wave naturally tends to move towards the west; but the obstructions offered by the continents and the promontories, and the irregular conformation of the bottom of the ocean, materially change the direction of its motion. Sometimes its direction is even towards the east. From a point about one thousand miles southwest of South America there appear to start two tidal waves, which travel in nearly opposite directions, one towards the west and the other towards the east.

178. *Four Daily Tides.*—At some places the tides rise and fall four times in each day. This is ascribed to the existence of two different tidal waves, coming from opposite directions. This phenomenon occurs on the eastern coast of Scotland, where one wave comes into the North Sea through the English Channel, while a second wave comes in around the northern extremity of

Scotland. At places where these two waves arrive at different times, each wave will produce two daily tides.

179. *Tides in Lakes and Inland Seas.*—If there is any tide in lakes and inland seas, it is usually so slight as to be scarcely measurable. A series of careful observations has demonstrated the existence of a tide in Lake Michigan, which is at its height about half an hour after the moon's transit. The average height which it attains, however, is less than two inches.

180. *Other Phenomena.*—Along the northern coast of the Gulf of Mexico there is only one tide in the day, the second one being probably obliterated by the interference of two waves. An approximation to this state of things is noticed on the Pacific coast, where at times one of the daily tides has a height of several feet, and the other a height of only a few inches. A very curious statement is made by missionaries in reference to the tides at the Society Islands. They say that the tides there are uniform, not only in the height which they attain, but in the time of ebb and flow, high tide occurring invariably at noon and at midnight: so that the natives distinguish the hour of the day by terms descriptive of the condition of the tide.

NOTE.—It is now generally admitted that one result of the friction of the tides is a diminution of the velocity of the earth's rotation; and it is possible that the moon's secular acceleration (page 133) is partly due, not to an increase in the moon's orbital velocity, but to this same diminution of the earth's rotation. The amount of the diminution is, however, so very small that all attempts to compute it have been thus far unsuccessful.

CHAPTER XII.

THE PLANETS AND THE PLANETOIDS. THE NEBULAR HYPOTHESIS.

181. *The Planets and their Apparent Motions.*—There are other celestial bodies besides the sun and the moon, which, while they share the common diurnal motion towards the west, appear to change their relative positions among the stars. These bodies are called *planets*, from a Greek word signifying wanderer. Some of them are visible to the naked eye, and some only become visible by the aid of a telescope. In some of them the change of position among the stars becomes apparent from the observations of a few nights: while in others even the annual change of position is so small that they have for ages been considered to be fixed stars.*

This change of position is determined, as it was determined in the case of the sun and the moon, by observations of their right ascensions and declinations. When such observations are made, the apparent motions of the planets are found to be very irregular. Sometimes they appear to move towards the east, and sometimes towards the west, while at other times they appear to be stationary in the heavens. Such irregularity in the direction of their motion is at once seen to be incompatible with the supposition

---

* The times of meridian passage of all the planets which ever become conspicuously visible are given in the American Ephemeris. The altitude which any planet has when on the meridian is obtained from the corresponding zenith distance, and this is found at any place whose latitude is known, from the formula

$$z = L - d:$$

the declination being also given in the Ephemeris. It must be remembered that $(L-d)$ becomes numerically $(L+d)$ when L and d have different names; that z in any case has the same name as $(L-d)$; and that when z is north, the bearing of the body is south, and conversely.

that they are, like the moon, satellites of the earth, revolving about it as a centre. The next supposition that will naturally be made is that they may revolve about the sun.

182. *Heliocentric Parallax.*—In order to test the correctness of this second supposition, we must first, from the apparent motions which are observed from the earth, deduce the corresponding motions which would be seen by an observer stationed at the sun. Fig. 64 will serve to show how, by means of a body's heliocentric parallax (Art. 56), the position which that body would have if seen from the sun's centre—in other words, its heliocentric position—may be determined from its geocentric position. In this figure let *S* represent the sun, *ABCE* the earth's orbit, the plane of which intersects the celestial sphere in the circle

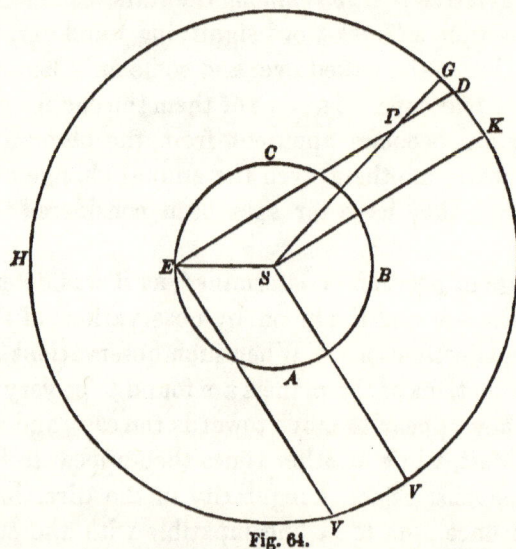

Fig. 64.

VHD, and *P* the position of a planet when projected upon the plane of the ecliptic. Let the vernal equinox be supposed to lie in the direction *EV* or *SV*, which two lines must be supposed sensibly to meet when prolonged to the celestial sphere. Draw the lines *EP* and *SP*. Since the distances of the planet from the sun and the earth are finite, these lines will not lie in the same direction, and the angle *EPS* which they make with each other will be the difference of the directions in which the planet is seen from the earth and the sun: in other words, the planet's heliocentric parallax in longitude. Through S draw SK parallel to ED. The angle VEP, or its equal, VSK, is the planet's geocentric longitude, and is obtained by observation. The sum of this angle and the angle EPS is the angle VSP, or the planet's heliocentric longitude. Provided, then, we know the angle EPS, we can readily obtain the angle VSP. Now, in the triangle PES we know from Kepler's Third Law the ratio of the sides SP and ES, which are the distances of the planet and the earth from the sun; and the angle PES, the planet's angular distance from the sun, or its elongation (Art. 56), can be obtained by observation. The angle EPS may then be readily computed.

By a similar method the planet's heliocentric latitude may be determined from its geocentric latitude, and the heliocentric place of a planet may thus be obtained at any time.

183. *Orbits of the Planets.*—When the motions of the planets as they would be seen from the centre of the sun are thus deduced from their observed motions with reference to the earth, all the apparent irregularities of motion disappear. The planets are found to revolve from west to east in ellipses about the sun in one of the foci, the eccentricity of the ellipses diminishing, as a general rule, as the magnitude increases. The planes of the orbits are found to be nearly coincident with the plane of the ecliptic, and Kepler's and Newton's laws are exactly fulfilled in the case of each planet. The line in which the plane of each planet intersects the plane of the ecliptic is called the line of nodes, and the terms perihelion and aphelion have the same signification that they have in the case of the earth.

184. *Inferior and Superior Planets.*—The planets are divided into two classes: inferior and superior planets. The *inferior* planets are those whose distances from the sun are less than the distance of the earth from the sun, and whose orbits are therefore included within the orbit of the earth. The inferior planets are Mercury and Venus.* The *superior* planets are those whose

---

* There are some astronomers who are inclined to suspect the existence of a third inferior planet, Vulcan, distant about 13,000,000 miles from the sun. Two American observers are convinced that they saw such a planet during the total solar eclipse of July 29, 1878.

distances from the sun are greater than that of the earth, and whose orbits therefore include the orbit of the earth. The superior planets are Mars, Jupiter, Saturn, Uranus, and Neptune. There is besides these a group of small planets, called *minor planets*, *planetoids*, or *asteroids*, situated between Mars and Jupiter. Up to October, 1878, 192 of these minor planets had been discovered. The earth is also a planet, lying between Venus and Mars. It is therefore a superior planet to Mercury and Venus, and an inferior planet to the other planets. Its sidereal period is greater than the periods of the inferior planets, and less than those of the superior planets.

INFERIOR PLANETS.

185. In Fig. 65 let S represent the sun, $PP'P''P''''$ the orbit of an inferior planet, the plane of which is supposed to coincide

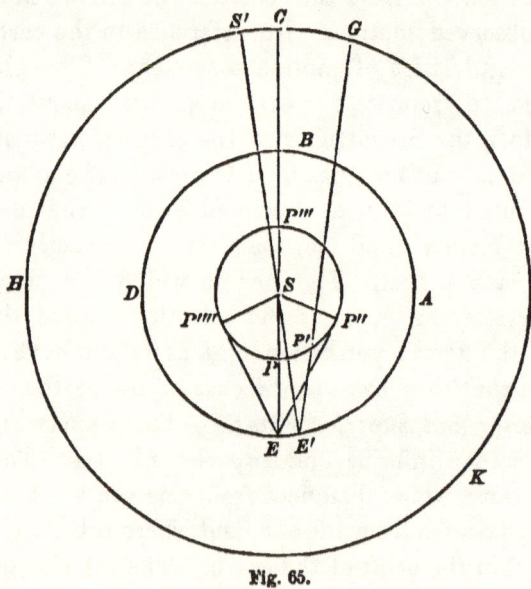

Fig. 65.

with the plane of the ecliptic, $ABDE$ the orbit of the earth, and the circle KGH the intersection of the plane of the ecliptic with the celestial sphere. Suppose the earth to be at E. When the planet is at P, between the earth and the sun, or at P''', on the opposite side of the sun to the earth, it has the same geocentric longitude as the sun, and is in conjunction with it. The position at P is called the *inferior* conjunction, and that at P'''' the *superior* conjunction.

The greatest angular distance of the planet from the sun will evidently occur when the line connecting the planet and the earth is tangent to the orbit of the planet; that is to say, when the planet is at P'' or P''''. The position at P'' is called the *greatest western* elongation of the planet, that at P'''' the *greatest eastern* elongation of the planet. We have already seen in Art. 92 that the relative distances of the planet and the earth from the sun are at once obtained when we have measured the greatest elongation.

186. *Direct and Retrograde Motion.*—The motion of the inferior planets is always in reality from west to east, or *direct*, as it is called; but when the planet is near its inferior conjunction, its motion is apparently from east to west, or *retrograde*. This apparent retrograde motion is explained in Fig. 65. Let the planet be at its inferior conjunction at P, and let both the earth and the planet move on about the sun in the direction $EABD$. The angular and the linear velocity of the planet about the sun being greater than they are in the case of the earth, when the earth arrives at E', the planet will be at some point P', and will lie in the direction $E'G$; the sun, on the other hand, will lie in the direction $E'S'$. While, then, the earth is advancing from E to E', the sun and the planet appear to move away in opposite directions from the point C, on which both were projected when the earth was at E. But the apparent motion of the sun is invariably towards the east; hence the planet has apparently moved towards the west. It may also be shown that the same apparent retrograde motion occurs when the planet is approaching inferior conjunction, and is within a short distance of it.

187. *Stationary Points.*—When the planet is at P'''', it is moving directly towards the earth in the direction $P''''E$, and the motion of the earth in its orbit gives the planet an apparent motion in advance. The same must also be the case when the planet is at P''. Since then the motion of the planet is direct at the greatest elongations, and retrograde at inferior conjunction, there must be a point in the orbit between inferior conjunction and each elongation at which the planet neither advances nor

recedes, but appears stationary in the heavens. These points are called the *stationary points*.

At all other parts of the orbit except those which have been discussed, the apparent motion of the planet is direct; but the velocity with which it moves is subject to great variation. It was on account of this irregularity, both in the direction and the amount of their apparent motions, that these bodies were called wandering stars by the ancient Greeks.

188. *Evening and Morning Stars.*—Except at the times of conjunction, an inferior planet is either to the east or to the west of the sun. When it is to the east of the sun it will set after the sun has set, and when it is to the west of the sun it will rise before the sun has risen. In certain parts of its orbit the planet's elongation from the sun is sufficiently great to carry the planet beyond the limits of twilight, or 18° (Art. 101); it will then be an *evening star* if to the east of the sun, and a *morning star* if to the west. It is only to Venus, however, that these terms are commonly applied, at least so far as the inferior planets are concerned: since Mercury is so near to the sun that it is seldom visible, and even when it is visible, it appears like a star of only the third or the fourth magnitude.

189. *Elements of a Planet's Orbit.*—In order to compute the position in space at any time of either an inferior or a superior planet, we must be able to determine:—

1st. The relative position of the plane of the planet's orbit to the plane of the ecliptic;

2d. The position of the orbit itself in the plane in which it lies;

3d. The magnitude and the form of the orbit; and

4th. The position of the planet in its orbit.

These four conditions require the knowledge of seven distinct quantities. The first condition is satisfied if we know (1) the position of the line in which the plane of the orbit intersects the plane of the ecliptic, or, what amounts to the same thing, the longitude of the planet's nodes, and (2) the inclination of the two planes to each other. The second condition is satisfied if we know (3) the longitude of the perihelion. The third condition is satisfied if we know (4) the semi-major axis, or the planet's mean distance from the sun, and (5) the eccentricity of the orbit. Finally, the last condition is satisfied if we know (6) the time in which it makes one complete revolution about the sun, or its *periodic time*, and (7) the time when the planet is at some known place in its orbit, as, for instance, the perihelion. These seven quantities are called the *elements* of the orbit.

190. *Heliocentric Longitude of the Node.*—A planet is at its nodes when its latitude is zero; and if the heliocentric longitude of the planet at that time can be determined, it will also be the heliocentric longitude of the node, since the line of nodes of every planet passes through the sun. But the heliocentric longitude of a planet when at its node differs from the geocentric longitude (which may be obtained directly from observation), excepting only in case the earth itself happens at that time to be on the line of nodes. We must therefore be able to deduce the heliocentric longitude of the planet from the geocentric. When the planet's distance from the sun is known, this can be done by the method explained in Art. 182; and when this distance is not known, the following method can be used.

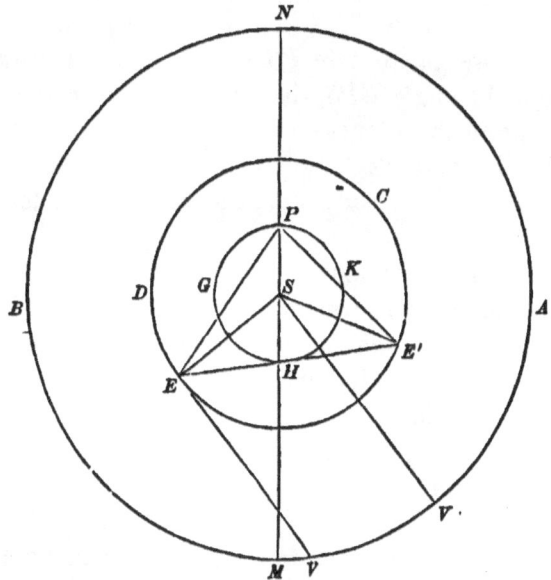

Fig. 66.

In Fig. 66, let S be the sun, $PGHK$ the orbit of a planet, $CDEE'$ the orbit of the earth, and NM the line of nodes of the planet. Let the vernal equinox lie in the direction EV or SV. Let E be the position of the earth when the planet is on the line of nodes at P. The elongation of the planet, or the angle PES, and the geocentric longitude of both planet and node, or the angle VEP, can be obtained by observation. Suppose both planet and earth to move on in their orbits, and the earth to be at E' when the planet again reaches the same node, and let the planet's elongation at this time, or the angle $SE'P$, be observed. Now, since the earth's orbit, although represented in the figure by a circle, is really an ellipse, ES and $E'S$ will not in general be equal to each other. The value of each, however, can be readily obtained from the solar tables. The same tables will also give us the angle ESE', which is the angular advance of the earth in its orbit in the interval. In the triangle ESE', then, knowing two sides and the included angle, we can compute the side EE', and the angles SEE' and $SE'E$. The angles PES and $PE'S$ having been obtained by observation, we can find the angles PEE' and $PE'E$. Then in the triangle PEE', knowing two angles and the included side, we can compute the side EP. Finally, in the triangle PES, knowing an angle and the two including sides, we can obtain PS, or the planet's distance from the sun, and the angle EPS, the planet's heliocentric parallax, from which, and the planet's observed geocentric longitude, we can obtain the planet's heliocentric longitude, as in Art. 182. This is, as we have already seen, the heliocentric longitude of the node.

The nodes of every planet are found to have a westward movement, similar in character to the precession of the equinoxes. It is a very slight movement, however, being only 70' a century in the case of Mercury, and being less than that in the case of the other planets.

191. *Inclination of the Planet's Orbit to the Ecliptic.*—The line of nodes of any planet being, as we have just now seen, a nearly stationary line in the plane of the ecliptic, the earth must pass it once in very nearly six months in its revolution about the sun. The inclination of the plane of any planet's orbit to the plane of the ecliptic may be determined by observations made when the earth is on the line of nodes. In Fig.

67 let E be the earth, and EN the line of nodes of a planet. Let AEN be the plane of the ecliptic, and P the position of a planet projected on the surface of the celestial sphere. With EP as a radius, let the arc PA be described, perpendicular to the plane of the ecliptic, and also the arcs PN and NA. In the spherical triangle PNA, right-angled at A, the arc PA, which measures the angle PEA, is the geocentric latitude of the planet; AN, or the angle AEN, is the difference between the geocentric longitudes of the planet and the node; and the angle PNA is the inclination of the plane of the orbit to the plane of the ecliptic. In the triangle PNA, we have,

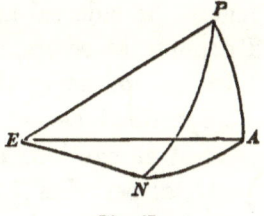

Fig. 67.

$$\tan PNA = \frac{\tan PA}{\sin AN}.$$

It may be noticed in this formula that, since the sine of a small angle varies more rapidly than the sine of a large angle, an error in AN will affect the result the less, the greater AN itself happens to be at the time of observation: that is to say, the farther the planet is from the node.

192. *The Periodic Time of an Inferior Planet.*—The time in which an inferior planet makes one complete revolution about the sun, or its *periodic time*, may be found by taking the interval of time between two successive passages of the planet through the same node. The accuracy of this method is however diminished by the small inclination (less than 7°) of the planes of the orbits to the plane of the ecliptic, which renders it difficult to determine the instant when the latitude of the planet is zero. It is found to be better to determine the planet's synodical period, or the interval between two successive conjunctions of the same kind, and from it to compute the sidereal period. The conditions of this problem, and the method by which it is solved, are identical with those in the case of the synodical and the sidereal period of the moon, Art. 141. The formula there given was,

$$P = \frac{ST}{S+T}.$$

In applying this to the case of an inferior planet, P and S

denote the sidereal and the synodical period of the planet, and T denotes, as before, the sidereal year.

Instead of using the interval between two conjunctions as the synodical period, we may take the interval between two greatest elongations of the same kind. A very accurate mean synodical period is obtained by taking two elongations separated by a long interval, and dividing this interval by the number of synodical periods which it contains.

193. It is hardly necessary to describe the methods by which the other elements given in Art. 189 are determined. It will readily be understood that the distance of the planet from the sun, obtained as in Art. 190, or by Kepler's Law, will enable us to obtain the form and the magnitude of the ellipse in which the planet moves. The method by which the longitude of the perihelion is obtained, although not intrinsically difficult, is too elaborate for this work. Finally, when the longitude of the perihelion is obtained, the time of the planet's perihelion passage may evidently at once be determined. The perihelion of Venus has a very minute retrograde motion: the perihelia of all the other planets have an eastward motion, similar to that of the earth's line of apsides.

MERCURY.

194. Mercury revolves about the sun at a mean distance of about 36,000,000 miles, the eccentricity of its orbit being about $\frac{1}{5}$th. Its synodical period is about 116 days, and its sidereal period 88 days. Its real diameter is about 3000 miles.* Its mass is a matter of considerable uncertainty, and quite different values of it are given by different astronomers. It may be assumed to be approximately equal to $\frac{1}{18}$th of the mass of the earth.

The greatest elongation of Mercury is only about $28° \ 20'$:

---

* The distances and the diameters of all the celestial bodies except the moon depend for their accuracy upon the accuracy with which the solar parallax is determined. An error of $1''$ in this parallax would affect the sun's distance from us to the amount of several millions of miles, and proportionally also the distances and the diameters of the other celestial bodies; the values given must therefore be considered to be only approximate.

and hence the planet is rarely visible to the naked eye, and is never a conspicuous object in the heavens. It is not generally believed that it has any atmosphere. It exhibits phases similar to those of the moon, and due to the same causes. It is asserted by some observers that it rotates on an axis in about 24 hours, though the truth of the assertion is by no means unchallenged. If it has any compression, it is extremely small.

VENUS.

195. The mean distance of Venus from the sun is about 67,000,000 miles, its synodical period is 584 days, and its sidereal period 225 days. The eccentricity of its orbit is small, being only about $\frac{7}{1000}$ths. Venus is nearly as large as the earth, its diameter being 7,600 miles. Its mass is about $\frac{4}{5}$ths of the mass of the earth. It has no perceptible compression.

The greatest elongation of Venus from the sun amounts to about 47° 15′, and hence it is often visible as an evening or a morning star. At certain times its brightness is so great that it can be seen in broad daylight with the naked eye, while at night shadows are cast by the objects which it illuminates. It exhibits phases similar to those of Mercury.

It seems to be generally admitted that Venus has an atmosphere the density of which is not very different from that of the earth's atmosphere. Observations of spots upon the disc go to show that the planet rotates upon an axis in a period of about $23\frac{1}{4}$ hours; but this conclusion is by no means certain.

The existence of a satellite of Venus was formerly suspected, but no satellite was seen at the transit in December, 1874.

196. *Transits of Venus.*—We have already seen (Art. 93) how a transit of Venus across the sun's disc is employed in determining the distance of the earth from the sun. If the plane of the orbit of Venus coincided with the plane of the ecliptic, a transit would occur whenever the planet came into inferior conjunction, or once in every 584 days. Owing to the inclination of the planes to each other, however, it is evident that at the time of inferior conjunction the planet may have too great a latitude to touch any part of the sun's disc. Now the phenomenon of a transit of Venus is analogous to a solar eclipse, and there-

fore if in Fig. 60 we suppose M to be Venus, the formula obtained in Art. 160 will apply equally well to the limits of the geocentric latitude of Venus within which a transit is possible. These limits are the sum of the semi-diameters of the sun and the planet and of the parallax of the planet, diminished by the parallax of the sun. The greatest value of the limits will be found to be about 17' 49''. When, therefore, the latitude of Venus is more than 17' 49'' at the time of inferior conjunction, no transit will occur: and as Venus in every sidereal revolution attains a latitude of over 3° 23', it is at once evident that a transit is only a rare occurrence.

197. *Intervals between Transits.*—Since the latitude of Venus is so small when a transit occurs, it is plain that the planet must be either at or very near one of its nodes. Now, let us suppose that Venus is at its node at the time of inferior conjunction, under which circumstances a transit will, of course, take place. The sidereal period of Venus is 224.7 days. Now, we have,

$$224.7d. \times 13 = 2921.11d.; \text{ and,}$$
$$365.256d. \times 8 = 2922.05d.$$

At the end of eight years, then, Venus will be very near the same node at the time of inferior conjunction, and a transit will probably occur. Again, we have,

$$224.7d. \times 382 = 85835.4d.$$
$$365.256d. \times 235 = 85835.16d.;$$

so that transits at the same node also occur every 235 years. In the same way transits may also occur at the other node: and the intervals between transits at either node are found to be 8, 105½, 8, 121½, 8, &c. years.* The longitude of the ascending node of Venus is 75° 20', and a transit at that node must occur, when it occurs at all, at the time when the earth is near that point, which is about the 6th of June. For a similar reason transits at the descending node occur about the 6th of December.

The last two transits occurred in June, 1769, and December, 1874. The next two will occur in December, 1882, and June, 2004. [See Table VIII., Appendix]

---

* Thus the years of transits at the ascending node are 1761, 1769, and 2004: at the descending node, 1639, 1874, and 1882.

SUPERIOR PLANETS.

198. The superior planets are, as they have already been defined to be, planets whose orbits include the orbit of the earth. They have, like the inferior planets, superior conjunction, but can evidently have no inferior conjunction. Their elongation from the sun, eastern and western, can have all values between 0° and 180°. When their elongation is 90°, they are said to be in *quadrature*, and when it is 180°, in *opposition*. Much that has already been said in this chapter in reference to the inferior planets is equally true in reference to the superior planets. The elements of the orbits are in both instances the same, and so are the methods by which the heliocentric longitudes of the nodes and the inclinations of the orbits are determined. In some other points there is a difference between the two classes of planets; and these points we shall now proceed to examine.

199. *Retrograde Motion.*—The superior planets, like the inferior planets, have at times an apparent retrograde motion, which

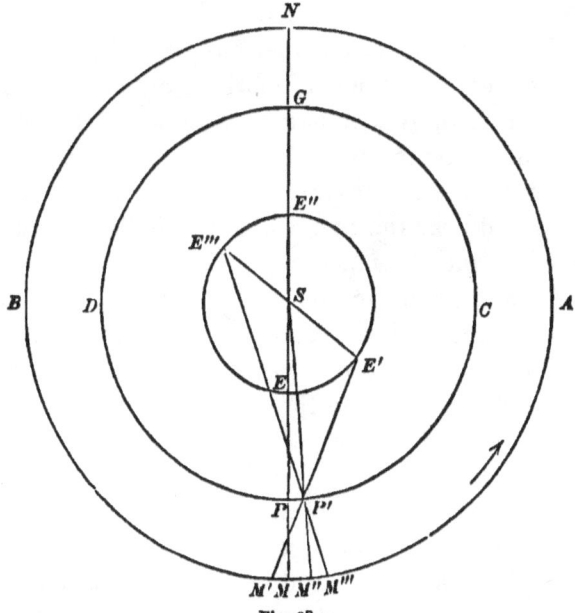

Fig. 68.

occurs at or near the time of opposition. The explanation of this retrogradation will be seen by a reference to Fig. 68. *S* is the sun, $EE'E''E'''$ the orbit of the earth, and $CGDP$ the orbit of a superior planet, the plane of which is supposed to coincide with the plane of the ecliptic, and to meet the celestial sphere in the circle $ANBM$. Let the earth be at E, and the planet at P, 180° in geocentric longitude from the sun. The planet will appear to be projected upon the celestial sphere at the point M. Let both earth and planet revolve in their orbits in the direction indicated by the arrow. When the earth has reached the point E', the planet, whose angular velocity is less than that of the earth, will have reached some point P', and will lie in the direction $E'M'$: in other words, it has apparently retrograded. If, on the other hand, the earth is at E'' and the planet at P, or in superior conjunction, the apparent motion of the planet is at that point direct, and its angular velocity appears to be greater than it really is; for when the earth is at E''', and the planet at P', the latter, having in reality moved through the arc MM'' since conjunction, appears to have moved through the arc MM'''.

The apparent motion, then, of a superior planet is direct in all cases except when it is at or near its opposition. The apparent motion of an inferior planet has been shown to be retrograde at and near the time of inferior conjunction (Art. 186). Now, since the earth is a superior planet to an inferior planet, and an inferior planet to a superior planet, we see, by comparing the two cases, that the retrograde motion occurs in each class of planets at and near the time when the inferior planet comes between the sun and the superior planet.

The stationary points in the orbit of a superior planet are identical in character with those in the orbit of an inferior planet, and occur when the retrograde motion is changing to the direct motion, or the direct to the retrograde.

200. *Synodical and Sidereal Periods.*—The synodical period of a superior planet is the interval of time between two successive conjunctions or two successive oppositions. When a planet is in conjunction with the sun, it is above the horizon only in the day time; but when it is in opposition, it is above the horizon during the night, and can therefore be readily observed. Hence in obtaining the synodical period it is better to employ the interval of time between two oppositions: and by determining the

times of two oppositions which are not consecutive, and dividing the interval between them by the number of synodical revolutions which it contains, we may obtain a mean value of the synodical period. By using times of opposition which were observed and recorded before the Christian era, a very accurate value of the mean synodical period may be obtained.

The method of deducing the periodic time from the synodical period is the same that was used in the case of the inferior planets (Art. 192), with the important exception that in the present case it is the earth that gains 360° upon the planet in the course of a synodical revolution, and not the planet that gains it upon the earth. If, therefore, we denote the periodic times of the earth and the planet by T and P, and the synodical period of the planet by S, we shall have (Art. 141),

$$\frac{360°}{T} - \frac{360°}{P} = \frac{360°}{S}$$

$$\therefore P = \frac{ST}{S-T};$$

which gives the value of any planet's sidereal period in terms of its synodical period and the sidereal year.

201. *Distance of a Superior Planet from the Sun.*—The distance of a superior planet from the sun may be obtained by the method of Art. 190: or it may be obtained from observations made at the time it is in opposition. In Fig. 69, let S be the sun, E the earth, and P a superior planet at the time of opposition, the planes of the two orbits being supposed to coincide. At the

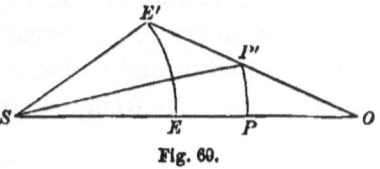

Fig. 69.

end of a short interval, let the earth have moved to E', and the planet to P'; the angle $E'OE$ will be the amount of the apparent retrogradation of the planet in that time. The periods of the earth and the planet being known, we can compute the angular advance of each planet in the given interval, thus obtaining the angles $E'SE$ and $P'SP$. The radius vector of the earth's orbit, SE', can be found from the Nautical Almanac. Then, in the triangle $E'SP'$, we know the side $E'S$, the angle $E'SP'$, which is the angular gain of the earth on the planet, and the angle $E'P'S$, which is the sum of $P'SO$, or the advance of the planet, and $P'OS$, or the apparent retrogradation of the planet. We can therefore compute the side SP', which is the distance required.

If we suppose the real distance of the sun from the earth not to be known, this method will still give us the ratio between this distance and the distance of the planet from the sun. And furthermore, if we can determine the real distance of the planet from the earth by observations of its parallax at the time of opposition (when it is nearest to the earth), by obtaining its displacement in right ascension when far east and far west of the same meridian, we can readily obtain the real distance of the earth from the sun. Such observations have been made upon the planet Mars, and the distance of the earth from the sun has been deduced, as already given (Art. 94).

202. *Evening and Morning Stars.*—The angular velocity of a superior planet towards the east is less than that of the earth, and consequently also less than the sun's apparent angular velocity in the same direction. After conjunction, therefore, the planet will lie to the west of the sun, and its elongation will continually increase. When this elongation exceeds about 30°, the planet will begin to be visible as a morning star, and will so continue until it has fallen 180° to the west of the sun, and is in opposition. It will then rise about sunset and set about sunrise. After the time of opposition it will lie more than 180° to the west of the sun, or, what is the same thing, less than 180° to the east of it, and will rise before sunset. It will therefore be an evening star from opposition to conjunction.

MARS.

203. The synodical period of Mars is 780 days, and its sidereal period 687 days. Its mean distance from the sun is 141,000,000 miles, and the eccentricity of its orbit about $\frac{1}{11}$th. Its diameter is 4,100 miles. Different values are given to its compression, varying between the limits of $\frac{1}{80}$th and $\frac{1}{18}$th. Its mass is about $\frac{1}{8}$th of that of the earth. It has two very small satellites, discovered by Prof. A. Hall, U. S. Navy, in 1877.

204. *Phases.*—At opposition and conjunction the same hemisphere is turned towards both the sun and the earth, and con-

sequently the planet appears full. At quadrature it appears slightly gibbous. It is the only one of the superior planets which exhibits any sensible phases, excepting possibly Jupiter.

Mars shines with a red light, and at opposition is a very conspicuous object, sometimes equalling Jupiter in brilliancy.

205. *Rotation, &c.*—When examined in a telescope, the surface of Mars is seen to be covered with patches of a dull reddish color, which are supposed to be land, interspersed with spots of a bluish or greenish hue, which are supposed to be water. By observation of these spots Mars is found to rotate upon an axis once in about $24\frac{1}{2}$ hours. The axis of rotation is inclined at an angle of 61° 18' to the plane of the ecliptic, and hence there must be a change of seasons not very different from the change which takes place on the earth. White spots are seen near the poles, which decrease in the Martial summer and increase in the winter. These spots are supposed to be snow. The existence of an atmosphere of a moderate density is generally admitted by astronomers.

THE MINOR PLANETS.

206. *Bode's Law.*—In 1778 the astronomer Bode, of Berlin, announced (though he did not discover) the following curious relation between the distances of the different planets from the sun. The statement of this relation usually goes by the name of "Bode's law." If we take the series of numbers

0, 3, 6, 12, 24, 48, 96, 192, 384,

each of which, except the second, is double the preceding one, and add 4 to each of these numbers, the resulting series,

4, 7, 10, 16, 28, 52, 100, 196, 388,

will approximately represent the relative distances of the planets from the sun.* Thus Mercury is 36,000,000 miles from the sun, and Venus 67,000,000 miles; and these distances are to each other nearly in the ratio of 4 to 7. There was however (the

---

* The last two numbers were not in the series as originally announced by Bode, since Uranus and Neptune had not then been discovered. The real distance of Neptune is one-fourth less than it should be, if this law were anything more than a coincidence.

minor planets being then undiscovered) a break in the series, there being no planet corresponding to the number 28; and Bode ventured to predict that another planet might be found to exist at that point of the series: that is to say, between Mars and Jupiter. A similar prediction was made by Kepler, about the beginning of the seventeenth century.

207. *Discovery of the Minor Planets.*—In 1800, six European astronomers formed an association for the express purpose of searching the heavens for new planets; and within the next six years four minor planets were discovered. These were named Ceres, Pallas, Juno, and Vesta. No more were discovered until the end of 1845, but since that time some have been discovered in nearly every year. The number discovered up to October, 1878 (including Ceres, &c.), was 192.

The mean distances of these bodies from the sun vary from 200,000,000 to 300,000,000 miles. They are all very small, the largest being probably not over 300 miles in diameter, and many of the others being too small to admit of measurement. Vesta is the only one which is ever visible to the naked eye, and its visibility is very rare. Some of the others are so small that they can scarcely be seen with the strongest telescope, even at opposition. Their total mass is about one-third of the earth's.

The French astronomer Leverrier has concluded that the mass of these minor planets is by no means sufficient to produce the perturbations in the orbit of Mars and in that of Jupiter which are believed to be due to the attractions of this group. It is therefore extremely probable that many other planets, hitherto undiscovered, belong to the same cluster.

208. *Olbers's Theory.*—Shortly after the discovery of the first four minor planets, Dr. Olbers advanced the theory that these planets were fragments of a single planet, which had been broken in pieces by volcanic action or by some other internal force. This theory is still in favor with many astronomers; others, however, object to it on the ground that if these bodies did formerly constitute one single body, their orbits ought to have a common point of intersection, which is very far from being the case. There are, however, certain striking resemblances in their orbits, which seem to argue something common in their origin.

JUPITER.

209. Jupiter is the largest planet of our system. At times it surpasses Venus in brilliancy, and even casts a shadow. It revolves about the sun at a mean distance of 481,000,000 miles. The eccentricity of its orbit is about $\frac{1}{20}$th. Its synodical period is 399 days, and its sidereal period 4333 days, or about 11.9 years. Its diameter is about 89,000 miles, and its volume is about 1400 times that of the earth. It rotates on an axis in a little less than ten hours, and has a compression of $\frac{1}{15}$th. Its phases are so slight as to be scarcely perceptible.

210. *Belts.*—When examined through a telescope, the disc of Jupiter is seen to be streaked with dark belts, lying nearly parallel to the plane of its equator. With powerful telescopes these belts are found to have a gray or brown tinge. They are sometimes nearly permanent for several months, and sometimes they change their shape materially in the course of a few minutes. There are usually one broad and several narrower belts on each side of Jupiter's equator.

It is generally supposed that Jupiter is surrounded by a dense atmosphere, and that these belts are fissures in this atmosphere, through which the dark body of the planet is seen. The distribution of the atmosphere in lines so nearly parallel to the equator is supposed to be due to currents in the atmosphere, similar in character to our trade-winds, but having a more decided easterly and westerly tendency, from the more rapid rotation and the greater size of the planet. A point on Jupiter's equator rotates with a velocity of about 28,000 miles an hour, while a point on our own equator rotates with a velocity of only about 24,000 miles a day.

211. *Satellites.*—Jupiter is attended by four satellites or moons, revolving about it from west to east. They are distinguished from each other by the numbers I., II., III., and IV., the first satellite being the nearest to Jupiter. The second satellite is about as large as our moon, and the others are somewhat larger. They are not usually visible to the naked eye, though a few instances to the contrary are on record. The distance of the first satellite from Jupiter is 270,000 miles, and that of the fourth is 1,200,000

miles. The first revolves about Jupiter in a period of 42 hours, and the fourth in a period of 16d. 17h.

212. *Phenomena Presented by the Satellites.*—The satellites, in the course of their revolution about their primary, present four distinct classes of phenomena, which are shown in Fig. 70. In this figure let S be the disc of the sun, and $EE'E''E'''$ the orbit

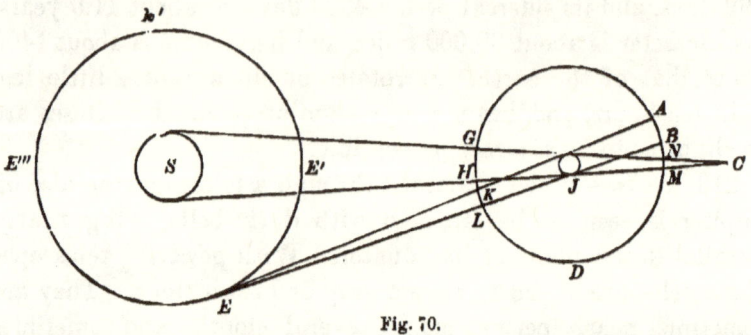

Fig. 70.

of the earth. Let J be Jupiter, and $ABDG$ the orbit of one of its satellites. Since the planes of all the orbits very nearly coincide with the plane of the ecliptic, we may consider $ABDG$ to lie in that plane. Suppose the earth to be at E, and, in order to simplify the case, suppose it also to remain at that point during the short time required by the satellite to revolve about Jupiter.

An *eclipse* of the satellite will occur when it passes through the arc MN, since it is then within the shadow formed by lines drawn tangent to the disc of the sun and that of Jupiter. It may readily be calculated that the length of the shadow, from J to C, is about 55,000,000 miles, so that the shadow extends far beyond the orbit of the fourth satellite. In extremely rare cases this satellite, owing to the inclination of its orbit to the ecliptic, may fail to be eclipsed.

An *occultation* of the satellite will occur when it passes through the arc AB, since it is then within the cone formed by lines drawn from E tangent to the disc of Jupiter.

A *transit of the shadow* will occur when the satellite passes through the arc GH, its shadow being then cast upon the disc of Jupiter, and moving across it as a small round spot.

Finally, a *transit of the satellite* will occur when it passes through the arc KL.

It will evidently depend on the relative situations of the sun, the earth, and Jupiter, whether all these phenomena will be observed or not. When the earth, for instance, is at E' or E'''', it is plain that only an occultation and a transit will occur.

The relative situations of the satellites to each other and to their primary are constantly changing. Sometimes all are on the same side of the primary: sometimes only one is visible, and sometimes, though very rarely, all are invisible. The times at which these different phenomena will occur are computed beforehand, and are given in the American Ephemeris, the time used being that of the meridian of Washington. The longitude of any place can therefore be obtained, at least approximately, by observations of these phenomena.

213. *Velocity of Light.*—If the transmission of light is not instantaneous, it is evident that the same phenomenon, if observed both at E' and E'''' (Fig. 70), will not occur at the same absolute instant of time at both places, but will occur later at E'''' by the time required for light to cross the orbit of the earth, a distance of 185,000,000 miles. And such is actually the case. This peculiarity was first noticed by Römer, a Danish astronomer, in 1675, who found that the times at which the phenomena occurred were earlier by about eight minutes at E', and later by the same amount at E'''', than the times computed for the mean distance of Jupiter from the sun. The time required by light in passing from E' to E'''' has been found by observation to be very nearly 16m. 27s.: whence the velocity of light is calculated to be 187,000 miles a second, a result agreeing very closely with the velocity obtained from the constant of aberration, discussed in Chapter VIII.

214. *Mass of Jupiter.*—The mass of Jupiter is much more accurately known than the mass of any of the planets which have hitherto been described. The reason is that Jupiter is attended by satellites whose distances from their primary, and whose periods of revolution, can be obtained by observation. We are thus enabled to compare directly the attraction which Jupiter exerts on one of its satellites with the attraction which

the sun exerts on Jupiter; and as, by the law of gravitation, the ratio of these two attractions is directly as the ratio of the masses of the two attracting bodies, and inversely as the square of the ratio of the distances through which these attractions are exerted, it is evidently within our power to obtain the ratio of the two masses. Since the attraction of the sun on Jupiter is equal to the centrifugal force of Jupiter in its orbit, if we denote the distance of Jupiter from the sun by D and its sidereal period by T, we have, by the formula of Art. 69, the expression for the sun's attraction on Jupiter equal to $\frac{4\pi^2 D}{T^2}$. In the same way, denoting the distance of a satellite from Jupiter by d, and its period by t, we have for the attraction of Jupiter on the satellite, $\frac{4\pi^2 d}{t^2}$: so that the ratio of the two attractions is $\frac{Dt^2}{dT^2}$. Finally, denoting the mass of the sun by M, and that of Jupiter by m, we shall have,

$$\frac{Dt^2}{dT^2} = \frac{M}{m} \times \frac{d^2}{D^2};$$

$$\therefore \frac{M}{m} = \frac{D^3 t^2}{d^3 T^2}.$$

By this formula an approximate value can be obtained of the mass of any planet which is attended by a satellite.

The mass of Jupiter is found to be $\frac{1}{1048}$th of that of the sun, or about 312 times that of the earth.

SATURN.

215. Saturn is, next to Jupiter, the largest planet of our system, and may fairly be considered to be the most interesting. It revolves about the sun at a mean distance of 881,000,000 miles, in an orbit whose eccentricity is about $\frac{1}{19}$th. Its synodical period is 378 days and its sidereal period 29.45 years. Its diameter is 72,000 miles, and it has a compression of about $\frac{1}{8}$th. It is attended by eight satellites, the planes of whose orbits, with one exception, very nearly coincide with the plane of its equator. Two of these satellites are rarely visible in any but the strongest

telescopes. Their distances from the planet range from 121,000 miles to 2,300,000 miles, and their sidereal periods from 22 hours to over 79 days. With one exception, they appear to be smaller than our moon.

216. *Rotation, &c.*—Saturn rotates upon an axis which is inclined at an angle of about 62° to the plane of the ecliptic, in a period of $10\frac{1}{4}$ hours. Belts are seen on the body of the planet, similar to those of Jupiter, although less marked. Other indications of the existence of an atmosphere have also been observed. The mass of the planet, determined by the motions of its satellites, is about $\frac{1}{3500}$th of that of the sun, or about 93 times that of the earth.

217. *Rings of Saturn.*—When observed through a telescope, Saturn is seen to be surrounded by a marvellous system of luminous rings, lying one within another in the plane of the planet's equator, and very nearly concentric with the planet. Although the planet itself was known to the ancients, the existence of these rings was not suspected until the seventeenth century. They were then supposed to be two rings, one within the other; but later observations, with improved instruments, have made it almost certain that the exterior of these two rings is itself composed of two, and it is probable that similar subdivisions also exist in the interior ring. A third independent ring, lying within the others, was discovered in 1850, by Professor Bond, at the Observatory of Harvard College.

Considering only the two rings which were first discovered, and neglecting the subdivisions that may exist in each, the approximate dimensions of the rings are as follows:—

Outer diameter of exterior ring..............	170,000 miles.
Breadth of exterior ring......................	10,000 miles.
Distance between the two rings................	2,000 miles.
Outer diameter of interior ring...............	146,000 miles.
Breadth of interior ring......................	17,000 miles.
Distance of interior ring from Saturn.........	20,000 miles.

218. The thickness of the rings is very small. Sir John Herschel estimated it at not more than 250 miles, while Professor Bond considered it to be only about 40 miles. The rings appear to rotate about the planet from west to east, the period of rotation

being about 10½ hours, according to Herschel. Various theories have been advanced as to their composition. The prevailing theory is that they are a collection of meteoric bodies, revolving about the planet precisely as similar rings or groups of meteoric bodies are found to revolve about the sun. (Chap. XIII.) A second theory is that they are composed of nebulous matter, and still a third is that they are solid. According to Professor Bond, however, they are in a fluid state; and there are many considerations which favor this view.

- The main cause of the stability of the rings in reference to the planet is undoubtedly the centrifugal force induced by the rapid rotation above mentioned. Professor Peirce, in developing the theory proposed by Professor Bond, maintains that the permanence of the equilibrium is finally dependent upon the attractions exerted by the satellites upon the rings. Herschel says that unless they were originally adjusted in their present position with the minutest precision, they must have been gradually formed under the influence of all the existing forces.

219. The rings must present a magnificent spectacle to the inhabitants of the planet. To an observer on Saturn's equator they will appear as an arch, passing through the zenith, and through the east and the west point of the horizon. To such an observer only the edge of the rings is visible. As he moves away from the equator, the altitude of the rings decreases, and the side of the rings becomes visible, presenting an appearance not unlike the familiar one of the rainbow. Under the most favorable circumstances of position, the rings will be projected against the sky as an arch with the enormous angular breadth of about $15°$, which is about 30 times the diameter which the sun presents to us.

220. *Disappearance of the Rings.*—As Saturn revolves about the sun, the plane of its rings remains, like the plane of the earth's equator, fixed in space, and intersects the plane of the ecliptic in a line which is called the line of nodes of the rings. In Fig. 71, let S be the sun, $ABCD$ the orbit of the earth, and $EHLN$ the orbit of Saturn. Let HN be the line of nodes of the rings, and draw the lines GO and KM parallel to HN, and tangent to the earth's orbit. When the planet is at H, the plane

of the rings passes through the sun, and only the edge of the rings is illuminated. In such a case the rings will disappear, or at all events will only be seen, in very powerful telescopes, as an exceedingly narrow line. Furthermore, if, while the planet is within the lines GO and KM, the earth encounters the plane

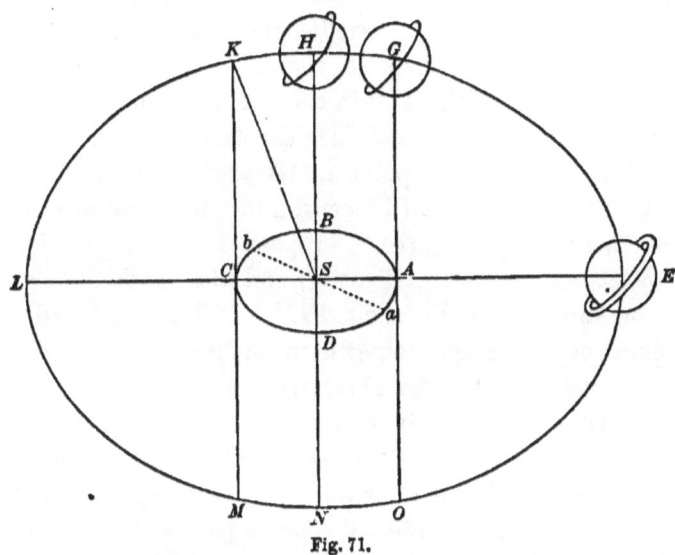

Fig. 71.

of the rings, they will again disappear. And thirdly, if, while the planet is within the same limits, the plane of the rings passes between the earth and the sun, the dark side of the rings will be turned towards the earth, and they will disappear. When the planet is beyond these limits, it is evident that the rings will always be visible, and will present an elliptical appearance, as represented at E.

Now, we can readily compute the length of time which Saturn requires in passing through the arc GK. For in the triangle CSK, right-angled at C, we know the sides CS and KS, or the distance of the earth from the sun and that of the planet, and can therefore obtain the angle CKS. It will be found to be about 6° 1'. This angle is equal to the angle KSH, and therefore double this angle, or 12° 2', is the angle through which Saturn moves about the sun in passing through the arc GK. Now we know that Saturn makes a complete revolution about the sun in

10,759 days; and therefore we may find by a simple proportion the time which it requires to pass through 12' 2'. This time is found to be 359.6 days, or very nearly a sidereal year; so that the earth makes very nearly one complete revolution about the sun while Saturn is passing through the arc GK.

221. *Number of Disappearances.*—Since Saturn's period of revolution is 29.45 years, these disappearances will occur at intervals of a little less than 15 years. Since the time during which the planet remains within the limits GO and KM is only six days less than a year, and since the earth may encounter the plane of the ring at any point in its orbit, it is quite certain that one such meeting will occur, and under certain circumstances there may be three. Suppose, for instance, that the earth is at a when Saturn is at G, and that both bodies move about the sun in the direction $EHLN$. The earth will meet the plane of the rings somewhere in the arc aA, and the rings will disappear. The rings will continue to be invisible for some time, since their plane will lie between the earth and the sun. The earth will overtake the plane before Saturn reaches the point H, and after that time the rings will be visible until the planet is at H, when the plane passes through the sun, and the rings again disappear. The earth will now be near the point b, and the rings will continue to be invisible, since their dark side is turned to the earth. The earth, passing through C, will again meet the plane somewhere in the arc CD, and after that time the rings will be visible. No more disappearances will occur for about fifteen years, at the end of which interval the planet will pass through the arc MO.

The last disappearance took place in 1878; the next will take place in 1892.

URANUS.

222. All the planets which have thus far been described, except of course the minor planets, were known to the ancients; but the last two are among the comparatively recent discoveries of astronomers. Uranus was discovered by Sir William Herschel, in 1781, by pure accident. Herschel, as well as other astronomers whose attention was directed to it,

at first supposed it to be a comet; and it was only after several months of observation that it was found to be a planet. Several names were suggested for it, but the name of Uranus was finally adopted. The astronomical symbol for it which the English have adopted is formed from the initial letter of Herschel's name.

Upon searching the star catalogues and other astronomical records, it was found that the planet had been observed no less than twenty times in the preceding 90 years, and had been considered to be a fixed star, its daily motion being so slight as to have escaped notice. Indeed its period is so great that even its annual change of position is only a few degrees.

223. These previous records, however, were of great assistance to astronomers in the determination of the elements of the planet's orbit. The synodical period of the planet is 370 days, and the sidereal period 30,687 days, or nearly 84 years. Its distance from the sun is 1,800,000,000 miles, and its diameter 33,000 miles. It is barely visible to the naked eye at opposition.

It is attended by four satellites, which are only visible in the most powerful telescopes; and by means of their movements the mass of the planet is found to be about 13 times that of the earth. One very remarkable point about these satellites is that their motion about their primary is *retrograde*, or from east to west, in planes inclined about 79° to the plane of the ecliptic, while the motions of all the other satellites which have hitherto been described are from west to east, and in planes making very small angles with the plane of the ecliptic. Sir Wm. Herschel believed that he had discovered four other satellites; but recent observations, with powerful telescopes, do not confirm his belief.

Scarcely more can be said of the physical appearance of Uranus than that it is uniformly bright. It exhibits neither spots nor belts, and therefore nothing can be determined as to any axial rotation, which is a question of special interest in the case of this planet, since one of the supports on which what is called "the nebular hypothesis" rests (Art. 228) is the assumption that the satellites of every planet revolve about it in the same direction in which it rotates on its axis.

NEPTUNE.

224. Early in the present century the conviction forced itself upon the minds of many astronomers that there must exist still another planet, exterior to Uranus. The circumstance which led to this conclusion was the existence of irregularities in the orbit of Uranus, over and above the irregularities which were due to the attractions exerted by the planets then known. The first systematic attempt to deduce the elements of the orbit of this unknown planet from these irregularities seems to have been made by Mr. Adams, of England, in 1843–5. The position which he assigned to the planet was in heliocentric longitude 329° 19′, but this determination was not then made public. The same intricate problem was also solved by M. Leverrier, of Paris, in 1845–6, and the longitude which he obtained was 326°. During the summer of 1846 search was made for the planet in England, but without success, owing to the want of a proper star-map. The observatory at Berlin, however, was better supplied; and on the night of September 23d, in compliance with a request made in a letter received that day from Leverrier, Dr. Galle at once detected, in longitude 326° 52′, what was apparently a star of the eighth magnitude, though it was not laid down on the map. Subsequent observations showed that this body was really a planet, and it was agreed to give it the name of Neptune.

"Such," in the words of Hind, "is a brief history of this most brilliant discovery, the grandest of which astronomy can boast, and an astonishing proof of the power of the human intellect."

225. The synodical period of Neptune is 367 days, and its sidereal period 60,127 days, or about 164 years. Its mean distance from the sun is 2,800,000,000 miles, and its diameter 37,000 miles. It is attended by one satellite, and some astronomers suspect the existence of a second. The mass of Neptune is about seventeen times that of the earth. The planet is not visible to the naked eye.

Nothing is yet determined as to the physical appearance or the axial rotation of the planet. A remarkable circumstance in connection with the satellite is that, like the satellites of Uranus, it moves about its primary *from east to west.*

226. It may help us in our conception of the immense distance of Neptune from us, even when it is in opposition, to consider that light, with its velocity of 186,000 miles a second, requires four hours to come from the planet to the earth. If there are any inhabitants of Neptune, the sun will to them have an apparent diameter of only $\frac{1}{30}$th of what it has to us, since the distance of Neptune from the sun is about thirty times that of the earth. It will therefore appear to them only about as large as Venus appears to us, under the most favorable circumstances. Saturn, Jupiter, and Uranus may possibly be visible to them as extremely small bodies, but it is very doubtful if any of the other planets of our system are visible, even with the strongest telescopes. We shall see hereafter, in the Chapter on Fixed Stars, that, with a base-line of even 185,000,000 miles (the diameter of the earth's orbit), attempts to determine the distances of the stars are, as a general rule, wholly unsuccessful; but it is very likely that similar attempts made by observers on Neptune, with a base-line thirty times as long as ours, may give more satisfactory results.

227. *Relative Sizes and Distances of the Planets.*—The relative distances of the planets from the sun, their relative magnitudes, as well as other numerical data concerning them, will be found in tables in the Appendix. In Plate I. will be seen a representation of their relative magnitudes, as they would appear to an observer stationed at the same distance from all of them.

THE NEBULAR HYPOTHESIS.

228. *Points of Resemblance in the Planetary Phenomena.*—The light of the planets and the satellites, when examined in the spectroscope, produces only the ordinary spectrum of reflected solar light. While, therefore, the spectral analysis of the light of the sun, the stars, and other heavenly bodies which shine by their own light, enables us to determine to some extent the elements of which they are composed, a similar experiment tells us nothing of the constitution of the planets or the satellites. What we do know, however, of their form, their appearance, their mass, and their density, leads us to conclude that they are bodies not dissimilar to the earth in general constitution. There are, besides, certain remarkable coincidences in the various phe-

nomena exhibited by the sun, the planets, and the satellites, which seem to point to a common origin of the whole solar system. The principal of these coincidences are the following:—

(1.) All the planets revolve about the sun in the same direction in which the sun rotates upon its axis: that is to say, from west to east.

(2.) The planes of the planetary orbits nearly coincide with the plane of the sun's equator.

(3.) The satellites of each planet, as far as known, revolve about their primary in the same direction in which the primary rotates upon its axis. The satellites of Uranus and Neptune may or may not form an exception to this rule, for these planets are so distant that observation fails to detect any axial rotation in them.

(4.) The planes of the orbits of the satellites of each planet approximately coincide with the plane of that planet's equator.

(5.) Both planets and satellites revolve in ellipses of small eccentricity.

229. *The Nebular Hypothesis.*—The idea of the nebular hypothesis seems to have presented itself at about the same time to both Sir William Herschel and Laplace. The principal points in it are the following. All the matter which now composes the sun, the planets, and the satellites once existed as a single nebulous mass, extending beyond the present orbit of Neptune, and rotating on an axis from west to east. In the progress of ages this nebulous mass slowly contracted and condensed, from the loss of the heat which it radiated into space, and from the gravitation of its particles towards the centre. As its dimensions became less, its velocity of rotation became greater, according to the laws of Mechanics: since any particle moving in a circle of any radius with a certain linear velocity would, as it approached the centre, move in a smaller circle with nearly the same linear velocity, and would therefore have a greater angular velocity. Finally, the centrifugal force generated by this increased velocity at the surface of the equator of the mass exceeded the attraction towards the centre, and a nebulous zone was detached, which revolved independently of the interior mass, just as the rings of Saturn have been seen to revolve about that planet. This

zone, by concentration at certain points within itself, broke up into separate masses; and these masses, either from slight differences of velocity or from the preponderating attraction of some fraction larger than the others, eventually formed one body, revolving about the central mass. And, furthermore, as these separate masses came together, a motion of rotation was communicated to the combined mass, just as a whirlpool or an eddy is formed when two streams of water meet; and this rotating mass, condensing and contracting in its turn, threw off from itself a second zone, which underwent all the changes above described. Thus were formed a planet and its satellite, each revolving about its primary in the direction of that primary's axial rotation: and by a continuation of the process the whole system of planets and satellites was evolved.

230. *Necessary Conditions.*—It is a necessary condition of the truth of this hypothesis, that the planets shall revolve (as they do revolve) about the sun in the same direction in which it rotates. It is also necessary that each satellite or system of satellites shall revolve about its primary in the same direction in which that primary rotates. It is not, however, absolutely necessary that the outer planets shall rotate in the same direction in which they revolve; although such a coincidence might be expected, since the revolution of the outer particles from which a planet was formed would be more rapid than that of those which were nearer to the sun.

If we assume this hypothesis to be true, the rings of Saturn are to be considered as rings which did not form satellites after they were thrown off from the planet; while in the case of the minor planets the ring broke up into separate masses, which have continued to revolve in independent orbits about the sun.

231. *Experiment in Support of the Hypothesis.*—The possible truth of the nebular hypothesis is supported by an ingenious experiment devised by M. Plateau:* A mass of olive-oil was immersed in a mixture of alcohol and water, the density of the mixture being made exactly equal to that of the oil. In this way

---

* See *Annales de Chimie*, vol. xxx. (1850). The experiment is also described in Carpenter's *Mechanical Philosophy*, &c., one of the volumes of Bohn's Scientific Library (London).

the mass of oil was practically withdrawn from the influence of gravitation. When made to rotate, the mass assumed a spheroidal form, and finally, when the velocity of rotation was sufficiently great, a ring of matter was thrown off in the equatorial region. This ring subsequently broke up into independent masses, each of which assumed a globular form, rotated on an axis of its own, and continued to revolve about the central mass: thus presenting precisely the successive phenomena which are assumed in the nebular hypothesis to have occurred in the formation of the solar system.

232. The truth of the nebular hypothesis is by no means universally admitted by astronomers and other scientific men; and it is difficult to say what is the predominant belief about it at the present time. The high scientific reputation of those who originated it, and of those who have since supported it, is sufficient justification for giving it a place in this treatise; but it must not be forgotten that its truth is still very emphatically an open question, and that many great minds are numbered with its opponents.

Sir William Herschel was led to the adoption of the nebular theory by his examination of that class of celestial bodies called nebulæ, some of which presented in his day, and present now, the appearance of masses of nebulous matter. Recent spectroscopic examinations of some of these nebulæ (Art. 286) go to show that they are really what they seem to be, masses of incandescent vapor; and this discovery gives a new interest to the nebular hypothesis. Mr. Lockyer, in his *Elementary Lessons in Astronomy*, says that "it may take long years to prove or disprove this hypothesis; but it is certain that the tendency of recent observations is to show its correctness."

Fresh doubts are thrown upon the truth of the nebular hypothesis by the discovery of the satellites of Mars (§ 203); since the angular velocity of the inner satellite appears to be three times as great as the rotation of the planet.

A statement of "Kirkwood's Law," which may have some bearing on the nebular hypothesis, will be found in the Appendix.

UNIV. OF
CALIFORNIA

PLATE III.

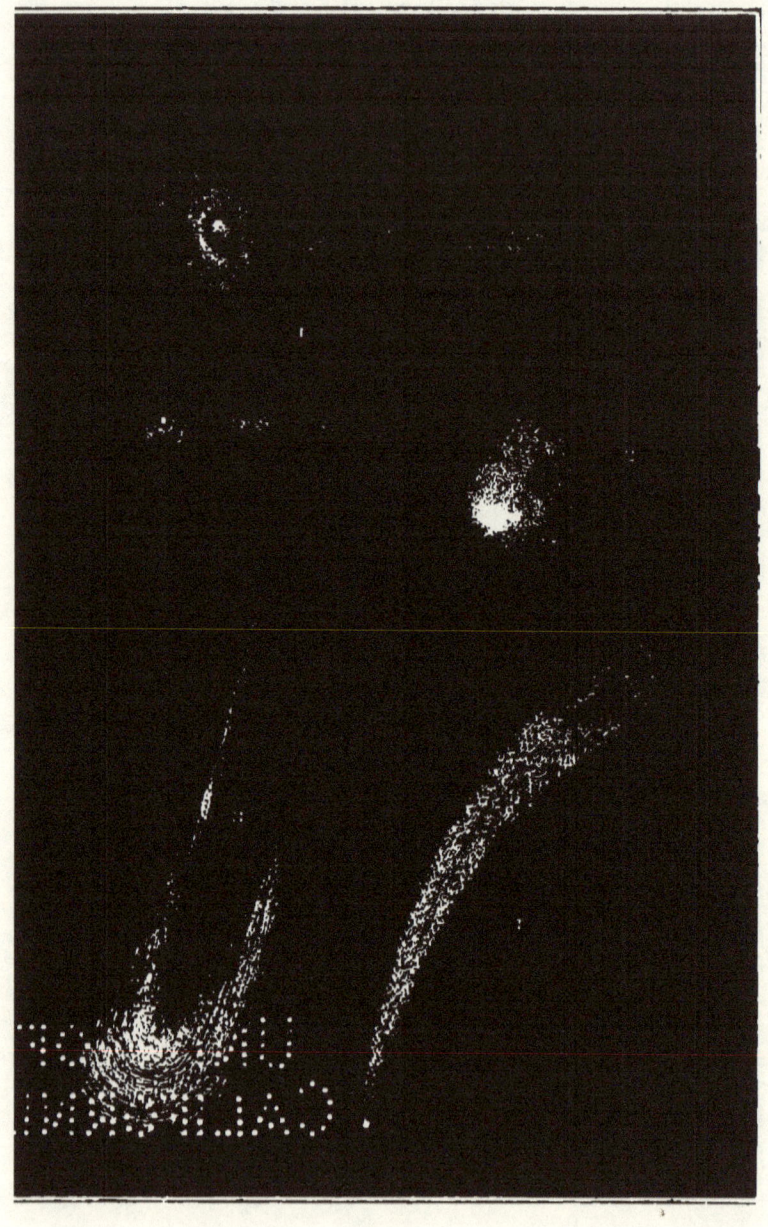

COMETS.

1. BIELA'S COMET
2. ENCKE'S COMET.
3. GREAT COMET OF 1861.
4. DONATI'S COMET, 1858.

CHAPTER XIII.

COMETS AND METEORIC BODIES.

COMETS.

233. *General Description of Comets.*—A comet is a body of nebulous appearance and irregular shape, revolving in an orbit about the sun. Comets have usually been considered to consist for the most part of nebulous matter; but the theory has lately been advanced that they are collections of minute meteoric bodies.

Comets differ widely from each other in appearance, and no description of them can be given to which there will not be many exceptions. Generally speaking, a comet consists of three parts: the *nucleus*, the *coma*, and the *tail*. The nucleus and the coma together form the *head*. The *nucleus* is a bright point, like a star or a planet, which may be either a solid mass, or a mass of nebulous matter of a density greater than that of the rest of the comet. The diameter of the nucleus varies considerably in different comets: that of the comet of 1845 (III)* was about 8000 miles, while that of the comet of 1806 was only 30 miles. The average value is not over 500 miles: and in many comets no nucleus whatever is perceptible.

The *coma* is a mass of cloud-like matter, more or less nearly globular in form, which surrounds the nucleus. The nucleus, however, as a general thing, is not situated at the centre of the coma, but lies towards that margin which is the nearer to the sun. The diameter of the coma is different in different comets: that of the comet of 1847 (v) was only 18,000 miles, while that of the comet of 1811 (I) was over 1,000,000 miles. Usually, however, it

---

* The number (III) means that this was the third comet which appeared in the course of the year.

is less than 100,000 miles. It is frequently noticed that the coma decreases in apparent diameter as the comet approaches the sun, and increases as the comet recedes from it. On the supposition that the coma consists of vaporous matter, this phenomenon is explained by the assumption that the intense heat to which the comet is subjected as it approaches the sun is sufficient to rarefy this vaporous matter to such an extent that some of it becomes invisible.

The *tail* is a train of cloud-like matter attached to the head, which usually lies in a direction nearly opposite to that in which the sun lies from the head. The tail is usually very small when the comet first appears, and sometimes is not even perceptible. As the comet approaches the sun, the length of the tail increases, and sometimes becomes enormous. In the comet of 1811 (I), for instance, the length of the tail was 100,000,000 miles; and in that of 1843 (I) it was 200,000,000 miles.

The *angular* length of the tail depends not only on its absolute length, but also on its distance from the earth, and on the direction in which the axis of the tail lies. There are six comets on record of which the tails subtended angles of over 90°; and one of these, that of 1861 (II), had a tail of 104° in length, as observed at some places.

234. *Diversity of Appearance.*—The description above given may be considered to apply to comets taken as a class; but, as already remarked, important exceptions are often noticed in individual comets. Indeed, it is hardly possible to compare any two comets without finding marked points of difference in them. Some comets are not visible at all, except by the aid of powerful telescopes, and are hence called *telescopic* comets; while others, again, are so conspicuous as to be visible to the naked eye in full daylight. Some comets have more than one tail; the comet of 1823, for instance, had a tail turned towards the sun, in addition to the usual one turned from it. The comet of 1744 is reported to have had six tails, spread out like an immense fan, through an angle of 117°; but the truth of the record is not above suspicion.

Not only do comets differ thus widely from each other in appearance, but even the same comet changes its appearance from day to day. Sometimes the nucleus decreases in diameter as it approaches the sun: sometimes its brightness increases, and jets of luminous matter are thrown off from it in the direction of the sun. The length of the tail often increases with marvellous rapidity; in the case of the Great Comet of 1843 (1), the increase was estimated to be about 35,000,000 miles a day, after the comet had passed its perihelion. There are some instances on record of a comet's having separated into two distinct comets. This is asserted in the Greek records of a comet which appeared in 370 B.C.: and Biela's comet presents an indubitable instance of this kind. This comet was observed in 1826 and 1832, and was determined to be a comet with a period of nearly seven years. Its return in 1839 was not observed. It again appeared in 1846, and then presented the extraordinary appearance of two comets, moving side by side, at a distance apart of over 150,000 miles.

235. *The Tail.*—The general form of the tail is that of a truncated cone, the larger base being at the extremity of the tail. It is noticed that the tail is always brighter near the borders than along the middle, from which it is inferred that it is hollow: since only on such a supposition would the line of sight pass through more luminous matter when directed to the edges than when directed to the middle. With regard to the formation of the tail, the most generally accepted theory seems to be that the matter of which the nucleus is composed is excited and dilated by the action of the sun's rays, as the comet approaches the sun, and that particles of vaporous matter are thrown off from it; and that these particles are driven to the rear by some repulsive force exerted by the sun, and thus form the tail. What this repulsive force exerted by the sun is, has not yet been determined; but the general situation which the tail of a comet has with reference to the sun seems to justify the inference that some such force does exist. Nor has it yet been determined what is the force which originally detaches these vaporous particles from the nucleus: it may be the same repelling force which drives them to the rear, it may be a force generated in the nucleus itself, or it may be a combination of both these forces. If we adopt the theory of the meteoric structure of these bodies, the tail is to be considered as a cloud of minute particles of matter, held together by their mu-

tual attraction, or by the attraction exerted upon them by the denser mass which constitutes the head.

236. *Curvature of the Tail.*—The tail of a comet is usually not straight, but is concave towards that part of space which the comet is leaving. If we assume the existence of a solar repulsive force, similar to that mentioned in the preceding article, this peculiarity of shape may be thus explained. In Fig. 72, let S be the sun, and GCD a portion of the orbit of a comet. When the nucleus is at A, let a particle be driven from it in the direction SA, with a force sufficient to carry it to L in the time in which the nucleus moves from A to C. When the nucleus reaches C, this particle, still retain-

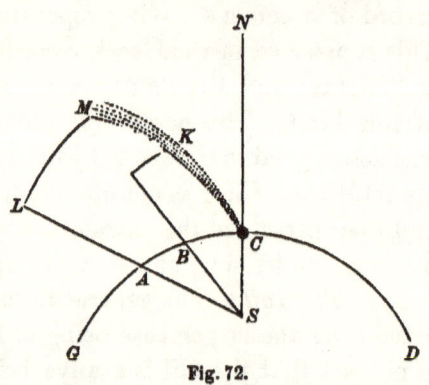

Fig. 72.

ing the motion which it had in common with the nucleus, will be found at some point M. In the same way a particle driven from the nucleus when it is at B will be found at some point K, when the nucleus reaches C: and, in general, when the nucleus is at C the tail will not lie in the direction SN, but in the direction of the curve CKM, as shown in the figure.

237. *Elements of a Comet's Orbit.*—A comet is identified at its successive returns, not by its appearance, which is liable, as we have already seen, to serious changes, but by the elements of its orbit. In consequence of the comparative ease with which the elements of a parabola can be calculated, astronomers are in the habit of using that curve to represent at first the approximate form of a comet's orbit. The elements of a parabolic orbit are five in number, and are as follows:—

(1.) The inclination of the orbit to the plane of the ecliptic:

(2.) The longitude of the ascending node:

(3.) The longitude of the perihelion:

(4.) The time at which the comet passes its perihelion:

(5.) The distance of the comet from the sun at perihelion:

Tables and formulæ have been constructed by which these

elements can be computed from the results of three distinct observations of the position of the comet: and these three observations may all be made, if necessary, within the space of 48 hours. The parabolic elements having thus been obtained, the catalogues of comets are searched to see if these elements are similar to those recorded of any previous comet. As it is highly improbable that the elements of any two comets will coincide throughout, the presumption is a strong one, if two comets, visible at different times, move in the same orbit, that they are one and the same comet: and the more often the coincidence is repeated, the more nearly does the presumption approach to a demonstration.

238. *Number of Comets, and their Orbits.*—The number of comets which have been recorded since the Christian era is over 770: and there are about 80 recorded as observed before that date. Of these 850 appearances of comets, some may undoubtedly have been only reappearances of the same comet: and, indeed, in some cases comets have been identified with other comets previously observed; but this can hardly be the case with the majority of these bodies. Besides these comets thus recorded, there must have been many others so situated as to be above the horizon only in the day-time: and such comets would become visible only in case of the occurrence of a total solar eclipse. A coincidence of this kind is recorded by Seneca as having occurred 62 B. C., when a large comet was seen in close proximity to the sun during a solar eclipse. The improvement of telescopes in recent years has greatly increased the number of comets which become visible, and 204 were observed between the years 1800 and 1876. We are justified, therefore, in concluding that the comets which have really come within our system since the Christian era are to be reckoned by thousands. Two centuries and more ago, Kepler made the remarkable statement that "there are more comets in the heavens than fishes in the ocean."

The orbit in which a comet moves may be either an ellipse, a parabola, or an hyperbola. The orbits of 329 comets have been subjected to mathematical investigations, and the results of these investigations may be thus tabulated: *

---

* The list of these comets, and the facts known concerning them, are given in G. F. Chambers's *Descriptive Astronomy* (Oxford, Eng.).

Comets with elliptical orbits 20;
Subsequent returns of these comets.................. 66;
Comets with elliptical orbits, which have not returned 43;
Comets with parabolic orbits.. .... 194;
Comets with hyperbolic orbits............................. 6.

A comet whose orbit is either a parabola or an hyperbola will not return to our system; provided, at least, that the attraction of other bodies does not alter the character of the orbit. It must be noticed, however, that some of the orbits which are called parabolic, may really be ellipses of an eccentricity so great as to render their elements undistinguishable from those of parabolas. In whatever conic section a comet may move, the sun is always at the focus.

239. *Periodic Times.*—The sixty-three comets which have been found to move in elliptical orbits differ widely from each other in the length of their periods. Among the twenty comets whose returns have been observed, there are seven with short periods, lying between three and fourteen years. There is no doubt that these seven comets are periodic; but there is some uncertainty with regard to some others of the remaining thirteen. Two elements of such uncertainty are the unsatisfactory character of the records of the observations made in the earlier ages, and the length of time which the periods embrace, being often several hundred years. Halley's comet, however, with a period of about seventy-five years, is unquestionably to be added to the list of periodic comets about which there is no doubt. There are five other comets, with periods not very different from that of Halley's, which have been discovered within the present century, and which have as yet made no return. With regard to the remaining comets to which elliptical orbits and periods have been assigned, little more can be said than that these periods embrace hundreds and even thousands of years.

240. *Motion of Comets in their Orbits.*—The motions of comets in their orbits about the sun are not performed in the same direction, the number of those whose motion is retrograde being about the same as the number of those whose motion is direct. According to Chambers, an examination of the motions of the various comets shows "that with comets revolving in elliptic

orbits there is a strong and decided tendency to direct motion. The same obtains with the hyperbolic orbits: with the parabolic orbits there is a rather large preponderance the other way; and taking all the calculated comets together, the numbers are too nearly equal to afford any indication of the existence of a general law governing the direction of motion."

The angles which the planes of the orbits make with the plane of the ecliptic have values ranging from 0° to 90°; but "there is a decided tendency in the periodic comets to revolve in orbits but little inclined to the plane of the ecliptic:" and if we take all the comets into consideration, "we find a decided disposition in the orbits to congregate in and around a plane inclined 50° to the ecliptic."

Owing to the great eccentricity of the orbits, some of the comets approach very near to the sun at the time of perihelion passage: the comet of 1843 (I), for instance, came within 100,000 miles of the sun's surface. For the same reason the distances to which some of the comets with elliptical orbits recede from the sun are immense; thus the comet of 1844 (II) receded to a distance of 400,000,000,000 miles, over 130 times the distance of Neptune from the sun. The velocity of the comets at perihelion is sometimes enormous; this same comet of 1843 swung about the sun through an arc of 180° in only two hours, and moved with the velocity of 350 miles a second.

241. *Mass and Density of the Comets.*—The minuteness of the mass of the comets is proved by the fact that they exert no perceptible influence on the motions of the planets or the satellites although they sometimes pass very near to them. Thus Lexell's comet, 1770 (I), in its advance towards the sun, became entangled with the satellites of Jupiter, and remained near them for five months, without sensibly affecting their motions. The effect of Jupiter's attraction on the comet, however, was very striking. The comet had a period of about five years, and yet it never appeared after 1770. It was found, by computation, that at its first return after that date it was so situated as not to become visible; and that in 1779, before its second return, it came nearer to Jupiter than Jupiter's fourth satellite: and the presumption is that its orbit was so changed and enlarged that the

comet no longer comes near enough to the earth to become visible. This same comet came within about 1,400,000 miles of the earth in 1770: near enough, had its mass been equal to that of the earth, to increase the length of the year by nearly three hours; but no sensible effect was produced.

The mass, then, of the comets being so small, and their volume so large, the density of the matter of which they are composed must be exceedingly rare. Indeed, it must be vastly more rare than that of the lightest gas or vapor of which we have any knowledge: for stars of the smallest magnitude are distinctly seen, and usually, too, with no perceptible diminution of brightness, through all parts of the comets excepting perhaps the nucleus; and this too in cases where the volume of nebulous matter has a diameter of 50,000 or 100,000 miles.

242. *Light of the Comets.* — The question whether or not comets shine by their own light does not seem to be satisfactorily decided. The existence of phases would of course prove that they shine by reflected light; but although in one or two cases the statement has been made that phases have been detected, the truth of the statement has in no case been universally accepted. Undoubtedly the distance of some comets is so great that phases might exist and still escape observation, as in the case of the superior planets; but, on the other hand, some comets come so near to the earth that there seems to be no good reason why phases, if any exist, should not be noticed. Observations upon the light of the comets have been made, both with the polariscope and with the spectroscope. The observations made with the polariscope seem to establish the fact that the comets shine, partly at all events, by the reflected light of the sun: as, for instance, the observations of Airy and others on Donati's comet in 1858. Mr. Huggins, of England, to whom we owe so many interesting discoveries made with the spectroscope, has recently examined the light of several comets with that instrument. In Brorsen's comet he found that the nucleus and part of the coma shone by their own light. In Tempel's comet the nucleus shone by its own light, and the coma by the reflected rays of the sun. In some comets the light from the nucleus resembled that which comes from the

gaseous nebulæ. In one comet there was a remarkable resemblance in the spectrum produced by its light to that produced by carbon, not only in the position of the bands, but in character and relative brightness.

Altogether, the question as to the light of the comets may fairly be regarded as still an open one, to be decided, perhaps, by future observations.

PERIODIC COMETS.

243. It has already been stated, in Art. 239, that there are at least eight comets which are undoubtedly periodic, seven of these being comets with short periods, and the eighth being Halley's comet. The following table contains a list of these comets.

NAME.	PERIOD IN YEARS.	NUMBER OF APPEARANCES OBSERVED.	LAST APPEARANCE.
Encke,	3.3	22	1878
Winnecke or Pons,	5.5	4	1875
Brorsen,	5.6	5	1879
Biela,	6.6	6	1852
D'Arrest,	6.6	4	1877
Faye,	7.4	5	1873
Méchain or Tuttle,	13.7	3	1871
Halley,	76.	5	1835

ENCKE'S COMET.

244. On November 26, 1818, a small and ill-defined telescopic comet was discovered in the constellation Pegasus, by the astronomer Pons, at Marseilles. It remained visible for seven weeks, and many observations were made upon it. Professor Encke, of Berlin, finding that the elements of the orbit did not agree with those of a parabola, determined to subject them to a rigorous investigation, according to the method proposed by

Gauss. This investigation showed that the orbit was elliptical, and that the period of the comet was about 3⅓ years. He further identified the comet with the comets of 1786 (I), 1795, and 1805, and predicted that it would return to perihelion on May 24, 1822, after being retarded about nine days by the influence of Jupiter.

"So completely were these calculations fulfilled, that astronomers universally attached the name of 'Encke' to the comet of 1819, not only as an acknowledgment of his diligence and success in the performance of some of the most intricate and laborious computations that occur in practical astronomy, but also to mark the epoch of the first detection of a comet of short period;—one of no ordinary importance in this department of science."

The comet has since been observed at every reappearance, the appearance in 1878 being the twenty-second on record. In 1835, it passed so near to the planet Mercury as to show conclusively that the generally received value of that planet's mass must be far too great: since the planet exerted no perceptible influence on the comet's orbit.

The comet is sometimes visible to the naked eye. It usually appears to have no tail; but in 1848 it had two, one about 1° in length, turned from the sun, and the other of a less length and turned towards it. At perihelion the comet passes within the orbit of Mercury: while at aphelion its distance from the sun is nearly equal to that of Jupiter.

One very curious feature in connection with this comet is that its period is steadily diminishing, by an amount of about 2½ hours in every revolution, the period having been nearly 1213 days in 1789–92, and only about 1210 days in 1862–65. Encke's own theory to account for this diminution is that the space through which the comet moves is filled with some extremely rare medium, too rare to obstruct the motions of the planets, but dense enough to offer sensible resistance to the progress of the comets. The effect of this diminution of velocity is to diminish the comet's centrifugal force, so that the comet is drawn nearer to the sun, and its orbit becomes smaller. But as the orbit becomes less, the angular velocity of the comet is increased, and its period of revolution is decreased.

This theory of Encke's concerning the existence of a resisting medium is by no means universally accepted by astronomers.

WINNECKE'S OR PONS'S COMET.

245. The second comet in the list was discovered by Pons, on June 12, 1819. Professor Encke assigned to it a period of $5\frac{1}{2}$ years, but the comet was not seen again until March 8, 1858, when it was detected by Winnecke, at Bonn. He was at first inclined to consider it a new comet, but soon identified it with the one previously discovered by Pons. Its distance from the sun at perihelion is about 70,000,000 miles, and its distance at aphelion 520,000,000 miles. It also appeared in 1869 and 1875.

BRORSEN'S COMET.

246. This comet was discovered by M. Brorsen, at Kiel, on February 26, 1846. The orbit was found to be elliptical, with a period of about $5\frac{1}{2}$ years, and its return to perihelion was fixed for September, 1851; but its position at that time was so unfavorable for observation that it was not detected. It was seen at its next return to perihelion, on March 29, 1857. It again escaped detection in 1862, but was seen in this country on May 11, 1868. It was seen also in 1873, and 1879.

Its perihelion distance is 60,000,000 miles, and its aphelion distance, 530,000,000 miles.

BIELA'S COMET.

247. This comet was discovered by M. Biela, an Austrian officer, at Josephstadt, Bohemia, on February 27, 1826. It was observed for nearly two months, and was identified with comets which had previously been seen in 1772 and 1805.

Its next return to perihelion was fixed for November 27, 1832; and the comet passed perihelion within twelve hours of that time. On October 29, 1832, it passed within 20,000 miles of the earth's orbit: but the earth did not reach that point of its orbit until a month afterwards. No little alarm was created, however, outside of the scientific world, when it became generally known how near to the earth's orbit the comet would approach.

At its return in 1839 it was not observed, owing to its close proximity to the sun. It was again detected on November 28, 1845, and by the end of the year it was found to have separated into two parts, and to present the extraordinary appearance of two comets, moving side by side, at a distance apart of over 150,000 miles. It again returned in 1852, and presented the same appearance; but the distance between the parts had increased to over one million of miles. Since that time the comet has never been seen, unless perhaps in 1872, as a meteoric shower.

Two theories have been advanced to account for this singular separation. One is that the division may have been the result of some internal repulsive force, similar to that which forms the tails of comets; the other is that it may have been the result of collision with some asteroid. At perihelion the comet passes within the orbit of the earth, and at aphelion it passes beyond that of Jupiter.

D'ARREST'S COMET.

248. This comet was discovered by Dr. D'Arrest, at Leipsic, on June 27, 1851. It remained in sight for about three months, and its period was determined to be about $6\frac{1}{2}$ years. Its return in November, 1857, was accordingly predicted, and the prediction was verified; although, owing to the comet's great southern declination, it was only observed at the Cape of Good Hope. The unfavorable situation of the comet in 1864 prevented its being seen; but it was seen on its returns in 1870 and 1877.

Its perihelion distance is about 100,000,000 miles, and its aphelion distance more than 500,000,000 miles.

FAYE'S COMET.

249. This comet was discovered by M. Faye, at the Paris Observatory, on November 22, 1843. It had a bright nucleus and a short tail, but was not visible to the naked eye. The elements of its orbit were investigated by Leverrier, who predicted that it would return to perihelion on April 3, 1851; and it returned within about a day of the time predicted. It has since made three returns, viz., in 1858, 1865, and 1873. The dimensions of its orbit are nearly the same as those of D'Arrest's comet.

MÉCHAIN'S OR TUTTLE'S COMET.

250. This comet was discovered by Méchain, at Paris, on January 9, 1790. Its period was calculated to be less than 14 years; but the comet was not seen again until January 4, 1858, when it was detected by Mr. H. P. Tuttle, at the Harvard College Observatory. Its third appearance was in 1871.

HALLEY'S COMET.

251. In the latter part of the seventeenth century, Sir Isaac Newton published his *Principia*. In that great work he assumed that the comets were analogous to the planets in their revolutions about the sun, although no periodic comet had then been discovered. He explained the methods of investigating the orbits of the comets, and invited astronomers to apply these methods to the various comets which had been observed. Halley, a young English astronomer, and afterwards the second Astronomer Royal, after a careful investigation, identified the comet of 1682 with comets which had appeared in 1531 and 1607: the period of the comet being about $75\frac{1}{2}$ years. The fact that the interval of time between the first and the second of these appearances was not exactly equal to that between the second and the third seemed at first to offer some difficulty; but Halley, "with a degree of sagacity which, considering the state of knowledge at the time, cannot fail to excite unqualified admiration," advanced the theory that the attractions of the planets would exert some influence on the orbits of the comets. Having thus decided that this comet was a periodic comet, Halley predicted the return of the comet about the beginning of the year 1759; and the comet passed its perihelion on March 12, in that year. The comet again appeared in 1835, and its next appearance will be in 1912.

The comet is a very conspicuous one, with a tail sometimes 30° in length and sometimes 50°. The comet has been traced back through the astronomical records, with more or less certainty, to 11 B.C., the number of appearances being about seventeen. It is not impossible that it was this comet which appeared in 1066, when it is recorded that a large comet excited dread throughout Europe, and was in England considered to presage the success of the Norman invasion. It is also probably identical with the comet of 1456, which had a splendid tail 60° in length.

Halley's comet at perihelion is nearer to the sun than Venus, while at aphelion it recedes beyond the orbit of Neptune.

REMARKABLE COMETS OF THE PRESENT CENTURY.

THE GREAT COMET OF 1811.

252. The comet of 1811 (I) was discovered on March 26, 1811, and was visible about seventeen months. It was very conspicuous in the autumn of 1811, remaining visible throughout the night for several weeks. Sir William Herschel states that the nucleus was well defined, with a diameter of about 428 miles; that it was of a ruddy hue, while the surrounding nebulous matter had a bluish-green tinge. The tail was about 25° in length, and 6° in breadth. Its aphelion distance from the sun is 14 times that of Neptune, and its period, according to Argelander, is 3065 years, with an uncertainty of 43 years.

THE GREAT COMET OF 1843.

253. The comet of 1843 (I) was first seen in the southern hemisphere in February, and became visible in the northern hemisphere the next month. It was decidedly the most wonderful comet of the present century. Its nucleus and coma shone with great splendor, and its tail was a luminous train of about 60° in length. On the day after its perihelion passage, and when only 4° distant from the sun, it was seen in broad daylight in some parts of New England, and its distance from the sun was measured with a sextant. It is described as having been at that time as well defined, in both nucleus and tail, as the moon is on a clear day. The comet is remarkable for its small perihelion distance, which was only about 540,000 miles; so that the comet came within 100,000 miles of the sun's surface. The intensity of the heat to which the comet must then have been subjected is almost inconceivable. Since 540,000 miles is about $\frac{1}{170}$th of the distance of the earth from the sun, and the inten-

sity of heat varies inversely as the square of the distance, the heat to which the comet was subjected must have been about 29,000 times as intense as the heat which prevails at the earth's surface: a heat nearly twenty times that required, as shown by experiments with powerful lenses, to melt agate and carnelian. For some days after this, the tail had a fiery red appearance; and its enormous length of over 200,000,000 miles, and the marvellous rapidity with which it was formed, were undoubtedly the results of the heat which it endured.

The rapidity with which this comet moved about the sun has already been noticed (Art. 240). The period has been computed to be about 175 years.

DONATI'S COMET.

254. This comet, 1858 (VI), was discovered on June 2d, by Dr. Donati, at Florence. It was then only discernible with a telescope, but became visible to the naked eye about the last of August. Indications of a tail began to be noticed about the 20th of August, and in a few weeks the tail assumed a noticeable curvature, which subsequently became one of the most interesting points connected with the comet. The comet passed its perihelion on September 29th, and was at its least distance from the earth on October 10th. Its tail subtended an angle of 60°, and had an absolute length of 51,000,000 miles. It disappeared from view in the northern hemisphere in October, but was seen in the southern hemisphere until March, 1859.

This comet was not as large as some others of the comets, but it was particularly noted for the intense brilliancy of its nucleus. The nebulosity surrounding the nucleus was also peculiar in its appearance. It consisted of seven luminous envelopes, parabolic in form, and separated from each other by spaces comparatively dark. These envelopes were detached in succession from the comet's nucleus, at intervals of from four to seven days. They receded from the nucleus with the daily rate of about 1000 miles. Perfectly straight rays of light, or "secondary tails," were also seen.

The comet has a period of about 2000 years. A magnificent memoir of this comet, by Professor G. P. Bond, is contained in the second volume of the Annals of the Harvard Observatory.

THE GREAT COMET OF 1861.

255. This comet, the second of the year, was discovered in the southern hemisphere on May 13th, but was not seen in England until June 29th, about two weeks after its perihelion passage. The nucleus was round and unusually bright, and the tail at one time attained the length of over 100°. The comet remained in sight for about a year.

"In a letter published at the time in one of the London papers, Mr. Hind, an English astronomer, stated that he thought it not only possible, but even probable, that in the course of Sunday, June 30th, the earth passed through the tail of the comet, at a distance of perhaps two-thirds of its length from the nucleus." Mr. Hind also stated that on Sunday evening there was noticed, by both himself and others, a peculiar illumination in the sky, like an auroral glare; and a similar phenomenon seems to have been noticed outside of London.

According to the observations of Father Secchi, the light of the tail, and that of the rays near the nucleus, presented evidences of polarization, while the nucleus itself at first presented no evidences whatever; afterwards, however, the nucleus presented decided indications of polarization. Secchi states that he thinks this "a fact of great importance, as it seems that the nucleus on the former days shone by its own light, perhaps by reason of the incandescence to which it had been brought by its close proximity to the sun."

METEORIC BODIES.

256. Under the general head of *meteors* are included three classes of bodies:—1st. The ordinary *shooting stars*, some of which can be seen rushing across the heavens on almost any clear night; 2d. *Detonating meteors*, which are shooting stars, commonly of an unusual size, whose disappearance is followed by a sound like that of an explosion; 3d. *Aerolites*, which, after the flash and the explosion with which they are generally accompanied, are precipitated to the earth in showers of stones and metallic substances. It is only within recent years that the decided attention of astronomers has been directed to these bodies, and comparatively

little is known with certainty about them; but the general belief is that they are all essentially of the same nature, differing from each other rather in size and density than in other more important respects.

257. *Shooting Stars.*—Scarcely a clear night passes during which shooting stars are not seen. The average number of those which can be seen at any place by one observer, on a cloudless moonless night, is estimated to be about six an hour. There is however, an hourly variation in the number observed, the minimum occurring about 6 P.M., and the maximum about 6 A.M. According to a French writer on this subject, the mean number of meteors observed is given in the following table:—

HOURS	7-8 P.M.	9-10	11-12	1-2 A.M.	3-4	5-6
MEAN NUMBER	3.5	4	5	6.4	7.8	8.2

It is further estimated that the number seen at any one place by a number of observers sufficient to watch the whole hemisphere of the heavens is 42 an hour, on the average, or about 1000 daily: and that the number which could be seen daily over the whole earth, under favorable circumstances, is more than 8,000,000. This is the number of those large enough to be visible to the naked eye: it is simply impossible to estimate the number of those which could be seen with the aid of telescopes.

It is further noticed that there are more shooting stars observed in the second half of the year than in the first. At certain seasons of the year, either in consecutive years or after the lapse of a certain number of years, there are unusually brilliant displays of these meteors, which are called *star showers*. The number of recognized star showers now exceeds fifty; and prominent among them are the shower of August 9-11 and that of November 11-13.

Professor Harkness, of the Washington Observatory, after an elaborate investigation of the quantity of matter in the ordinary shooting star, concludes that it is not far from *one grain*.

258. *The November Shower.*—There are several historical notices of brilliant displays of meteors which occurred in the early centuries of the Christian era: and ten of these, occurring between the years 902 and 1698, took place in October or November. The first display, however, of which we have any detailed account, occurred in 1799, on the morning of the 13th of November, and was visible over nearly the whole of the western continent. Humboldt witnessed it in South America, and thus describes it:—"Towards the morning of the 13th we witnessed a most extraordinary scene of shooting meteors. Thousands of bodies and falling stars succeeded each other during four hours. Their direction was very regular, from north to south. From the beginning of the phenomenon there was not a space in the firmament equal in extent to three diameters of the moon which was not filled every instant with bodies or falling stars. All the meteors left luminous traces or phosphorescent bands behind them, which lasted seven or eight seconds."

Similar showers also occurred on the same day of the month in the years 1831, 1832, and 1833, the last one being the most splendid on record. It lasted from ten o'clock on the night of the 12th to seven o'clock on the morning of the 13th, and was visible over nearly the whole of North America. The display reached its maximum about four A.M. An observer at Boston about six o'clock counted 650 shooting stars in a quarter of an hour. Large fireballs with luminous trains were also seen, some of which remained visible for several minutes. Even stationary masses of luminous matter are said to have been seen: and one in particular is mentioned as having remained for some time in the zenith over the Falls of Niagara, emitting radiant streams of light.

The November shower was witnessed again in 1866, both in this country and in Europe; but the display was much more brilliant in Europe. The maximum seems to have taken place about two A.M. on the 14th, when nearly 5000 meteors were counted in an hour at Greenwich. At half-past one, 124 were counted in one minute. The case was reversed with the shower of 1867, the display being more brilliant in this country than in Europe. The report on the shower from the United States Observatory at Washington states that as many as 3000 were counted in one hour. The most magnificent phase seems to have occurred about half-past four A.M. on the 14th. Professor Loomis states that at New Haven about 220 a minute were counted at this time. Many others were undoubtedly rendered invisible by the light of the moon, which was then very nearly full. Most of the brighter meteors left trains of phosphorescent light, which remained visible for several seconds, and in some cases for several minutes.

In 1868, the display began somewhat before midnight on the 13th and continued until daybreak on the 14th. Professor Eastman, of the Washington Observatory, says in his report, that "considering the number and brilliancy of the meteors, their magnificent trains, and the magnitude of the meteoric group through which the earth passed, this shower was unquestionably the grandest that has ever been witnessed at this Observatory." Over 5000 meteors were counted, and it was estimated that at five A.M. on the 14th the number falling in the whole heavens was about 2500 an hour. Several very brilliant meteors were observed. One in particular was brighter than Jupiter. It was at first of a deep orange color, afterwards green, and finally light blue. It left a train of 7° in length, which passed through the same changes of color, and remained visible for half an hour. The paths of 90 meteors were traced upon a chart, and were found in nearly every instance to start from a point in the constellation Leo.

259. *Height, &c. of the Meteors.*—Concurrent observations were made at Washington and Richmond, in November, 1867, for the purpose of determining the parallax of the meteors, and thence their distance. It was found that they appeared at an average height of 75 miles, and disappeared at the height of 55 miles. The velocity with which they moved relatively to the earth was 44 miles a second. Other observations have given nearly the same results.

The light of the meteors is probably due to the intense heat generated by the resistance of the air to the progress of these bodies. Notwithstanding the extreme rarity of the air at the height of the meteors, it is still believed that the heat resulting from such immense velocity is sufficient to fuse any known substance. A body moving with this velocity at the earth's surface would acquire a temperature of at least 3,000,000°. An examination of the light of the meteors with the aid of the spectroscope, by Mr. A. Herschel, showed that some of the meteors were solid bodies in a state of ignition, but that most of them were gaseous.

260. *Orbits of the Meteors.*—It is noticed that the November meteors, or at all events the great majority of them, seem to come from the same point in the heavens,—a point in the constellation Leo. So also the August meteors come from a point near the head of the constellation Perseus. Such points are called *radiant points*. Other showers have also other radiant points, situated in various parts of the heavens. The number of such points now recognized is more than 60. The paths in which meteors having the same radiant point move during the instant of time that we see them, are really parallel straight lines, the apparent convergence of the paths being merely the result of perspective; in other words, the radiant point is the vanishing point (Art. 16) of these parallel lines.

Knowing the direction and the velocity with respect to the earth of the motion of a meteor, it is easy to compute the same elements of its motion with reference to the sun. The results of such computation, together with the existence of the radiant points and the periodic recurrence of showers, have led to the theory that the November meteors are collected in a ring, or in several rings, or possibly in a series of clusters or groups, which revolve about the sun; and that the showers occur when the earth encounters these rings or groups. The theory of one ring is exemplified in Fig. 73.* Let $ABCD$ represent the orbit of the earth, and $AGBE$ a ring of meteors revolving about the sun. If A and B are the points at which the earth enters the ring, there will be displays of meteors, when the earth is at these points, similar to the August and the November shower. Unless the plane of the ring coincides with the plane of the ecliptic, there will be no showers at any other points of the earth's orbit. If we imagine the ring to be broken, or to be of unequal

---

* Phipson's *Meteors, Aerolites, and Falling Stars.* London, 1867.

thickness, or to consist of a series of groups, we may account for the irregularities which exist in the annual showers; and by

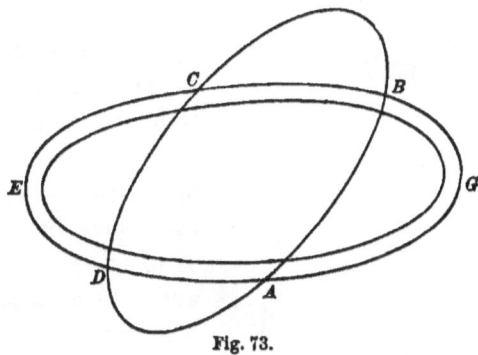

Fig. 73.

supposing the ring to have a period of about 33¼ years, we may account for the extraordinary displays of meteors which happen in about that interval of time. The duration of a shower, and the known velocity of the earth in its orbit, enable us to obtain an approximate value of the breadth of the ring. Professor Eastman estimates that the breadth of that portion of the ring through which the earth passed in November, 1868, could not have been less than 115,000 miles. The breadth of the stream in 1867 was less than this, but more densely packed with meteors.

Leverrier, a French astronomer, has computed the elements of the orbit of the November meteors. He finds the major semi-axis to be 10.34, the perihelion distance 0.989 (the radius of the earth's orbit being unity), and the eccentricity 0.9044. This would carry the aphelion beyond the orbit of Uranus, if both orbits were projected upon the plane of the ecliptic.

According to Professor Loomis, the relative situations of the orbit of the November meteors and the orbits of the earth and the other planets, are represented in Fig. 74; the orbit of the meteors being a very eccentric ellipse, the aphelion of which lies beyond the orbit of Uranus, and the period of the meteors being 33¼ years. According to the same authority, the August meteors revolve in a similar but much more eccentric ellipse, of which the aphelion lies far beyond the orbit of Neptune. The theory has also been advanced that meteors (or, at all events, some of them) are to be regarded as satellites of the earth rather than of the sun. On this subject Sir John Herschel says, in his *Outlines of Astronomy:*—"It is by no means inconceivable that the earth, approaching to such as differ but little from it in direction and velocity, may have attached them to it as permanent satellites, and of these there *may* be some so large as to shine by reflected light, and to become visible for a brief moment; suffering, after that, extinction by plunging into the earth's shadow."

261. *Detonating Meteors.*—The height and the velocity of these bodies are not essentially different from those of the ordinary shooting stars. They are, however, generally of an unusual brilliancy, and their appearance is followed by an explosion, or a series of explosions, the intensity of which is sometimes terrific. Records of more than eight hundred detonating meteors are to be found in scientific journals. The phenomena connected with the appearance of these bodies are, however, so nearly identical in character, that one instance may suffice to exemplify all. "On the 2d of August, 1860, about 10 P.M., a magnificent fireball was seen throughout the whole region from Pittsburg to New Orleans, and from Charleston to St. Louis, an area of 900 miles in diameter. Several observers described it as equal in size to the full moon, and just before its disappearance it broke into several fragments. A few minutes after the flash of the meteor there was heard throughout several counties of Kentucky and Tennessee a tremendous explosion, like the sound of

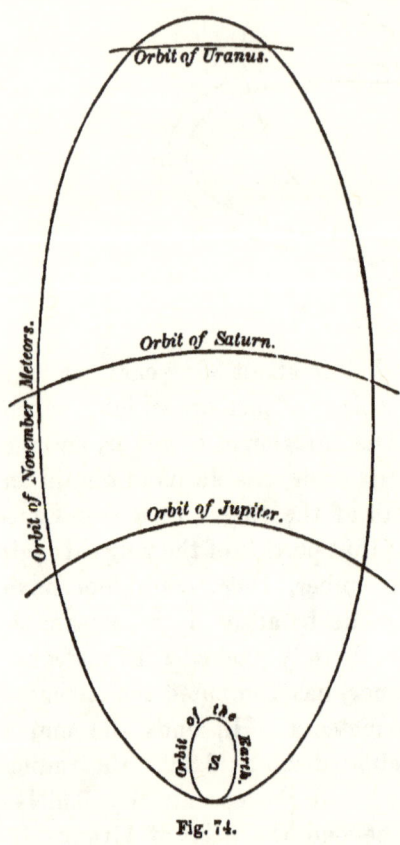

Fig. 74.

distant cannon. Immediately another noise was heard, not quite so loud, and the sounds were re-echoed with the prolonged roar of thunder. From a comparison of a large number of observations, it has been computed that this meteor first became visible over Northeastern Georgia, about 82 miles above the earth's surface, and that it exploded over the southern boundary line of Kentucky, at an elevation of 28 miles. The length of its visible path was about 240 miles, and its time of flight eight seconds: showing a velocity relative to the earth of 30 miles per second. It is hence computed that its velocity relative to the sun was 24 miles per second."

The explosions are probably due to the sudden compression and shocks to which the air is subjected as the meteor rushes through it, as happens when a gun is fired; or to the rushing of the air into the vacuum which the body creates in its rear. The appearance of these bodies is so sudden, and their velocity so great, that it is almost impossible to obtain any definite value of their magnitude. The diameters of some of them are stated to have been several thousand feet in length, but the estimate must be taken with considerable caution, particularly as it is impracticable to distinguish between the meteor itself and the blaze of light which surrounds it.

262. *Aerolites.*—Although the ordinary shooting stars sometimes appear to break in pieces, there is no evidence that any part of them falls to the earth. But occasionally solid masses of stone or of metallic substances do fall to the earth, their fall being usually preceded by the flash and the discharge of a detonating meteor. There is no doubt whatever about the authenticity of most of these cases, and the record of them extends far back into ancient history. A fall of meteoric stones near Rome, 650 years before Christ, is mentioned by the historian Livy; and a large block of stone is said to have fallen in Thrace, near what is now called the Strait of Dardanelles, 465 years before Christ. The entire number of aerolites of which we have any determinate knowledge is more than 400; and more than twenty falls of aerolites have occurred in the United States since the beginning of the present century. The British Museum contains a large collection of aerolites, one of which

weighs 8287 pounds; and many other similar specimens are to be found in the cabinets of colleges and museums, both in this country and in Europe. The following are instances of falls which have occurred since 1800.

In 1807, on the morning of December 14th, a brilliant meteor, with an apparent diameter equal to about one-half of that of the moon, was seen moving over the town of Weston, in the southwestern part of Connecticut. After its disappearance, three loud explosions were heard, followed by a continuous rumbling. Fragments of stone were precipitated to the earth within an area of a few miles in diameter. The entire weight of these fragments is estimated to be about 300 pounds. One fragment, weighing 36 pounds, is preserved in the museum at Yale College. The specific gravity of the aerolite was about $3\frac{1}{2}$, and among its components were silex, oxide of iron, magnesia, nickel, and sulphur.

On the 1st of May, 1860, about noon, there was a number of explosions over the southeastern part of Ohio. Stones were seen to fall to the earth, and in some cases they penetrated the earth to a distance of three feet. About thirty fragments were found, the largest of which weighs 103 pounds, and is to be found in the cabinet of Marietta College. The combined weight of these thirty fragments is not far from 700 pounds, and the specific gravity and the composition are very similar to those of the Weston aerolite.

Another phenomenon of this character occurred in Piedmont, on February 29, 1868, about the middle of the forenoon. There was a heavy discharge like that of artillery, followed, after a short interval, by a second discharge. A mass of irregular shape was seen in the air, enveloped in smoke, and followed by a long train of smoke. Other bodies, similar in appearance to meteors, were also seen. The analysis of the stones which fell showed the existence in them of the components mentioned above, and also of copper, manganese, and potassium.

Other aerolites have been subjected to chemical analysis. Of the 65 elementary substances known, 24 at least have been found in aerolites, and no new elements have been discovered. A meteoric shower usually consists of meteoric iron and meteoric stone; the iron is an alloy of which the principal part is nickel, and which also contains cobalt, tin, copper, manganese, and carbon; the stone contains chiefly those minerals which are abundant in lava and trap-rock. The proportions in which these ingredients enter into the composition of different aerolites differ greatly: sometimes an aerolite contains 96 per cent. of iron, sometimes scarcely any iron at all. A substance called *schreibersite*, which is a compound of iron, nickel, and phosphorus, is always found in these bodies.

The explosion of an aerolite may be due either to the intense heat generated by its rapid motion, or to the pressure to which it is subjected by the resistance of the atmosphere.

263. *Origin of Aerolites.*—Many theories have been advanced to explain the origin of these bodies. One theory is that they may be formed in the atmosphere by the aggregation of minute particles drawn up from the surface of the earth; but one objection to this theory is that it does not account for the nearly horizontal direction in which they move, and for the great velocity of their motion. A second theory, that they are thrown from terrestrial volcanoes, is open to the same objection. A third theory, that they may be ejected from the volcanoes of the moon, is weakened by the fact that observation shows no signs (or at least almost no signs) of activity in the lunar volcanoes. The most probable theory is that they are, like the planets and the comets, satellites of the sun, revolving about it in orbits which intersect the orbit of the earth, and that their fall to the earth's surface is either the direct result of their own motion, or is due to the resistance of the atmosphere, and the attraction exerted upon them by the earth.

264. *Possible Connection of Comets and Meteoric Bodies.*—The facts which have been presented in this chapter in relation to comets and meteoric bodies point to one certain conclusion:— that space, or at least that portion of space through which the earth moves, must be considered to be filled with a countless number of comparatively minute bodies, the aggregate mass of which cannot fail to be very great. In 1848, Dr. Mayer, of Germany, advanced a theory that the light and the heat of the sun are caused by the incessant collision of meteoric bodies with

its surface. In connection with this subject, Professor William Thompson states that if the earth were to fall into the sun, the amount of heat generated by the shock would be equal to that which the sun now gives out in 95 years; and that the planet Jupiter, under similar circumstances, would generate an amount of heat equal to that given out by the sun in 32,000 years.

There is a striking similarity between the elements of the orbit of Tempel's comet and those of the orbit of the November shower, mentioned in Art. 260. There is also a similar coincidence in the orbit of the August shower and that of the Great Comet of 1862; and the opinion is gaining ground with astronomers, not only that each of these comets leads the group with the elements of whose orbit its own elements so nearly coincide, but also that there is a close connection, generally, between comets and meteoric bodies. The following statement of this new theory is taken from an article by Professor Simon Newcomb, in the North American Review of July, 1868.

"The planetary spaces are crowded with immense numbers of bodies which move around the sun in all kinds of erratic orbits, and which are too minute to be seen with the most powerful telescopes.

"If one of these bodies is so large and firm that it passes through the atmosphere and reaches the earth without being dissipated, we have an aerolite.

"If the body is so small or so fusible as to be dissipated in the upper regions of the atmosphere, we have a shooting star.

"A crowd of such bodies sufficiently dense to be seen in the sunlight constitutes a comet.

"A group less dense will be entirely invisible, unless the earth happens to pass through it, when we shall have a meteoric shower."

It is not impossible to conceive that the planets themselves may have been formed by the aggregation of these minute bodies, a method of formation exactly the opposite of that which is set forth in the nebular hypothesis. An article in which such an origin is suggested for the planets, the comets, and Saturn's rings, will be found in the North American Review for 1864.

CHAPTER XIV.

THE FIXED STARS. NEBULÆ. MOTION OF THE SOLAR SYSTEM.

265. We have now examined the motions and the orbits of all the known members of the solar system. We have seen that the planets and their satellites, besides their apparent diurnal motion towards the west in orbits whose planes are perpendicular to the axis of the celestial sphere, have also independent motions of their own in elliptical orbits, the sun being at the common focus of the orbits of the planets, and each planet being at the common focus of the orbits of its satellites. Besides these bodies, there is a vast number of other bodies visible in the heavens, the phenomena presented by which are radically different from those which have hitherto been noticed. Continuous observations have been made upon the stars from year to year, and even from century to century; and it has been found that, after the results of these observations have been freed from the effects of precession and nutation (which, by shifting the position of the points or the planes of reference, may give the stars an apparent motion), the real change of position of the stars is extremely small. Sirius, the brightest star in the whole heavens, has an annual motion of $1''$; α Centauri, the brightest star in the southern hemisphere, has an annual motion of nearly $4''$; 61 Cygni and ϵ Indi, both small stars, have annual motions of $5''$ and $7''$ respectively. In only about 30 stars, however, has the amount of this change of position been found to be greater than $1''$ a year; and in the others, if any motion at all is detected, it is only that of a few seconds in a century. These motions are called *proper* motions, to distinguish them from those which are only apparent, and the stars are called *fixed* stars: a term which must be understood to imply, not that they have no motion in space, but that whatever motion they

have makes no perceptible alteration in their position upon the celestial sphere.

When a star moves obliquely to the line joining the earth and the star, its motion in its orbit can be resolved into two motions: one along the line of sight, either directly towards the earth or directly from it, and the other at right angles to that line. This latter motion will cause the star to shift its apparent position upon the celestial sphere, and will be the proper motion above described; or, as it may be called, the *transverse* proper motion. Now, it is evident that in order to obtain the real motion of any star in space we must be able to determine, not only its transverse motion, but also its motion towards or from the earth. As long as the detection of such a motion depended upon our ability to detect either an increase or a diminution of the star's brightness, its immense distance from us rendered the task a hopeless one; but very recently the spectroscope has afforded us the means of solving this problem. The details of this method will be given at the end of this Chapter.

266. *The Number of the Fixed Stars.*—The number of stars in the entire sphere which are visible to the naked eye is between 6000 and 7000, according to the largest estimate; but the number of those visible at any one time at any place is less than 3000. By means of telescopes, thousands and even millions of other stars are brought into view, in such numbers as almost to defy any attempt at computation.

267. *Magnitudes.*—The fixed stars are classified arbitrarily by astronomers according to their relative brightness, the different classes receiving the name of *magnitudes*, and the first magnitude comprising those stars which are the brightest. Different astronomers, however, sometimes assign different magnitudes to the same star. According to Argelander's classification, there are 20 stars of the first magnitude, 65 of the second, 190 of the third, 425 of the fourth, &c., the numbers in the following magnitudes increasing very rapidly. It is estimated that there are at least 20,000,000 stars in the first fourteen magnitudes. Those stars which are visible to the naked eye are comprised in the first six magnitudes; and stars of the twentieth magnitude are detected with the most powerful telescopes.

There is, however, a great difference in the brightness of stars which belong to the same magnitude. Sirius, for instance, is fifteen times as bright as some other stars of the first magnitude.*

268. *Constellations.*—In order to facilitate the formation of catalogues of the stars, they are separated into groups, called *constellations.* Ptolemy, in the second century, enumerated 48 constellations: 21 northern, 12 zodiacal, and 15 southern. The twelve zodiacal constellations have the same names that the signs of the zodiac bear, which are given in Art. 91; indeed, the signs really took their names from the constellations. Owing, however, to the precession of the equinoxes, the signs and the constellations no longer coincide (see Art. 119), the constellation of Aries being in the sign of Taurus, &c.

Since the time of Ptolemy, about sixty other constellations have been added to the list. Not all of these, however, are accepted by astronomers, and the list of those constellations which are generally acknowledged comprises only about 86 : 29 northern, 12 zodiacal, and 45 southern.

269. *Remarkable Constellations.*—The most remarkable of the northern constellations is that called Ursa Major, or the Great Bear, often called "The Dipper" from the well-known appearance presented by its seven conspicuous stars. The two of these seven

---

* Any one of the most prominent stars may be identified when on the meridian, in the following manner. The right ascension of the star, given in the Ephemeris, is, by Art. 9, the local sidereal time of the star's transit; and from this sidereal time the local mean solar time can be obtained, as proved in Art. 105, by subtracting from it the right ascension of the mean sun. When only a rough estimate is desired, the quantity given on page II. of each month in the Ephemeris, in the last column on the right, may be taken as the mean sun's right ascension. This will give the time of transit within four minutes. The more rigorous process is to apply to the quantity taken from page II. a correction taken from Table III. in the Appendix to the Ephemeris, using the longitude of the place as an argument, and adding the correction if the longitude is west. The result is then subtracted from the sidereal time as above, and the remainder is diminished by a correction taken from Table II., with the remainder itself as an argument. (See example in the latter part of the Ephemeris.)

The star's meridian altitude is found by the method given in the note to Art. 181.

stars which are the most remote from the handle of the dipper are called *the pointers*, since the right line joining them will always, when prolonged, pass very nearly through the pole-star. This constellation contains fifty-three stars of the first five magnitudes, including one of the first and three of the second.

There is another "Dipper," much less conspicuous, consisting also of seven stars, the pole-star being at the extremity of the handle. These stars form a part of the constellation Ursa Minor, which contains, in all, twenty-three stars of the first five magnitudes. The pole-star itself is of the second magnitude.

The constellation Orion is a magnificent one, and was fancifully supposed by the ancients to bear some resemblance to a giant. It contains two stars of the first magnitude, four of the second, and thirty-one of the next three magnitudes. The three stars, situated nearly in a straight line, which form the giant's belt, are a very conspicuous part of the constellation. This constellation, in the northern hemisphere, bears south about 9 P.M. in the early part of February; its altitude at that time at any place being nearly equal to the co-latitude of that place. Sirius, the brightest star in the heavens, is also seen at that time to the left of this constellation, and a little below it.

The constellation Pegasus contains forty-three stars of the first five magnitudes. Four of these stars are of the second magnitude, and nearly form a square. This square bears south about 9 P.M. in the early part of October, with an altitude, in the northern hemisphere, about 15° greater than the co-latitude of the place of observation.

The constellation Gemini, or the Twins, takes its name from two bright stars, nearly of the first magnitude, called Castor and Pollux. They are situated near each other, and bear south about 9 P.M. in the latter part of February, in the northern hemisphere, their altitude being about 30° greater than the co-latitude of the place of observation.

270. *Stars of the Same Constellation.*—Stars of the same constellation are distinguished from each other by the letters of the Greek or the Roman alphabet, or by numerals. The Greek letters were first used by Bayer, a German astronomer, in 1604, who called the brightest star in a constellation a, the next

brightest, β, &c. Thus the pole-star bears the astronomical name of α Ursæ Minoris, and the pointers of the dipper are called α and β Ursæ Majoris. Owing, however, either to carelessness on the part of Bayer or to changes in the brightness of some of the stars, this alphabetical arrangement does not in all cases accurately represent the relative brilliancy of the stars in a constellation.

The entire number of stars now catalogued amounts to several hundreds of thousands. Three catalogues published by Argelander, in 1859-62, contain over 320,000 stars observed at Bonn. In large catalogues, the stars are usually numbered from beginning to end, in the order of their right ascensions.

271. *Stars with Special Names.*—Some of the stars, particularly the more conspicuous ones, have special names, which were given

*α	Eridani	Achernar.
α	Tauri	Aldebaran.
α	Aurigæ	Capella.
β	Orionis	Rigel.
*α	Argus	Canopus.
α	Canis Majoris	Sirius.
α	Canis Minoris	Procyon.
β	Geminorum	Pollux.
α	Leonis	Regulus.
α	Virginis	Spica.
α	Boötis	Arcturus.
α	Scorpii	Antares.
α	Lyræ	Vega.
α	Aquilæ	Altair.
α	Piscis Australis	Fomalhaut.

to them by ancient astronomers. Instances are given in the accompanying table, all the stars contained in it being commonly considered to be of the first magnitude. About 90 stars are thus named, though most of these names are no longer used.

All of these stars, excepting those marked with an asterisk, come above the horizon throughout the United States. The position of each, in right ascension and declination, can be found in the Ephemeris for any day in the year. Besides the stars in the preceding table, there are three others of the first magnitude: α Crucis, α Centauri, and β Centauri. They are all stars of large southern declination, and do not come above the horizon in any part of the United States excepting the southern parts of Texas and Florida.

272. *Constitution and Diversity of Brightness.*—The spectra of stars are found to contain dark lines, similar in character to those by which we have seen that the solar spectrum is distinguished (Art. 102). These systems of lines differ from the system of lines in the solar spectrum, and they are also different in different stars. This difference, however, consists in the absence of certain lines seen in the solar spectrum, and not in the presence of new ones: nearly every line observed having its counterpart in the solar spectrum. The examination of these spectra, and the comparison of the dark lines which they contain with the bright lines found in the spectra of terrestrial substances, enable us to establish the presence of certain of these substances in the stars, precisely as was done in the case of the sun. In Aldebaran, for instance, the presence of sodium, magnesium, tellurium, calcium, antimony, iron, bismuth, and mercury, has been detected; in Sirius, of sodium, magnesium, iron, and hydrogen; in α Orionis, of magnesium, sodium, calcium, and bismuth.

The diversity of brightness in the stars may be due either to a difference in their distances from us, or to a difference in their actual magnitudes, or to a difference in the intrinsic splendor with which they shine. Probably all these causes exist; but it is fair to conclude that, as a general rule, the brightest stars are the nearest to us. Observations made for the purpose of determining the distances of the stars go to justify such a conclusion; although they also show that the rule is not an absolute one, since some of the fainter stars are found to be nearer to us than some of the brighter ones.

273. *Distance of the Fixed Stars.*—No perceptible difference

DISTANCE OF THE STARS.

is detected in the position of a star when observed at places of widely different latitudes. The conclusion drawn from this fact is, that the stars are so distant that lines drawn from any two points on the earth's surface to the same star are sensibly parallel: in other words, that the stars have no geocentric parallax. To determine the distance of the stars, then, we must have recourse to their heliocentric parallax. In Fig. 75 let S be the sun, $AE'BE$ the orbit of the earth, s the position of a fixed

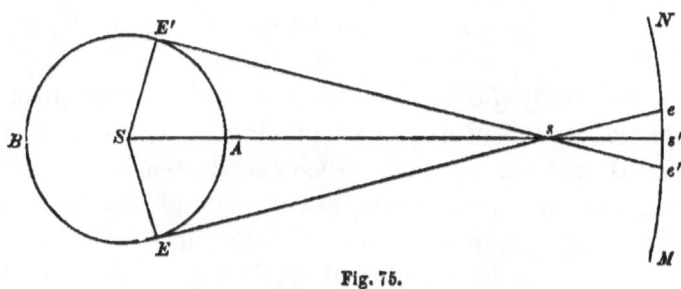

Fig. 75.

star, supposed to lie in the plane of the ecliptic, and NM a portion of the celestial sphere. From s draw lines sE, sE', tangent to the earth's orbit, and also prolong them beyond s, until they meet the arc of the celestial sphere NM. Draw the radii vectores SE' and SE to the points of tangency. If we suppose the star to be at rest, it will lie in the direction $E's$, when the earth is at E', and in the direction Es, when the earth is at E, and the motion of the earth about the sun will give the star an apparent oscillatory movement over the arc ee'. The true heliocentric direction in which the star lies is Ss'; and the difference of the directions in which the star lies at any time from the sun and the earth is its heliocentric parallax. This difference of direction is evidently at its maximum when the earth is at E' or E; and this maximum, or the angle SsE, is called the annual heliocentric parallax, or simply the *annual parallax*.

Numerous attempts have been made to determine the annual parallax of the stars, by comparing observations made when the earth is at E and E'. The nicest observation, however, has failed to detect in any star a parallax as great as 1″; and in only 12 stars has the slightest appreciable parallax been discovered.

274. The distance of the fixed stars, then, is so great that the radius of the earth's orbit, 92,400,000 miles, does not subtend an angle of even 1″ at that distance. If, in the triangle SsE, we suppose the angle SsE to be equal to 1″, we shall have,

$$Ss = 92,400,000 \text{ cosec } 1'':$$

which will be found to be about *nineteen trillions* of miles. This is only the inferior limit of the distance of the stars: that is to say, whatever the distance may be, it cannot be less than this; but how much greater it may be, particularly in the case of those numerous stars in which no movements whatever of parallax can be detected, it is impossible to calculate.

275. *Immensity of this Distance.*—It is hardly possible to obtain a clear conception of a distance of nineteen trillions of miles. Perhaps the nearest approach to such a conception is made by considering that light, moving with a velocity of 186,000 miles a second, and passing from the sun to the earth in a little more than eight minutes, would consume about $3\frac{1}{3}$ years in accomplishing such a distance: so that when we look at the brightest stars in the heavens, we see them, not as they are now, but as they were $3\frac{1}{3}$ years ago; and if any one of them were to be destroyed at any instant, we should continue to see its image for three years and more after that time.

Before such distances, the dimensions of the solar system shrink to the insignificance of a mere point in space. Neptune, the most distant of the planets, is nearly three billions of miles from the sun; and yet, if Neptune and the sun could both be seen from the nearest fixed star, the angular distance between them would never be greater than about 30″, which is only about $\frac{1}{60}$th of the angle which the sun subtends to us.

276. *Differential Observations.*—In dealing with so small a quantity as the annual parallax of the stars, it is important to avoid all circumstances by which even the most minute errors may be entailed upon the observations. The apparent position of a star is affected, not only by parallax, but by precession, nutation, aberration, and the star's own proper motion. The laws of precession and nutation enable us to decide what amount of the apparent motion is due to them; and the effect of a star's proper motion is also readily separated from that of parallax,

DISTANCE OF THE STARS. 221

since the former changes the position of the star from year to year, while the latter only changes its position *during the year*, causing it to lie now on one side and now on the other of its true position, but giving it no annual progressive motion. But it is not so easy to separate the effects of aberration and parallax. Aberration, as we have already seen (Art. 125), causes a star to describe a circle, an ellipse, or an arc, about its true position as a centre, according to its situation with reference to the plane of the ecliptic; and it is easy to see that the parallactic movement of a star is of precisely the same character. Thus we have already seen, in Fig. 75, that a star situated in the plane of the ecliptic will oscillate by parallax over the arc *ee'*; and if the star is not in the plane of the ecliptic, lines drawn from all points of the earth's orbit to the star, and thence prolonged to the celestial sphere, will evidently meet the sphere in a circle if the star is at the pole of the ecliptic, and in an ellipse if it is not at the pole.

In order to separate the effects of parallax and aberration, the following method was adopted by the astronomer Bessel. Instead of attempting to determine by direct observation the change of position of the star whose parallax was sought, he selected another star of much less magnitude, and therefore supposed to be at a much greater distance, which lay very nearly in the same direction as the first star, and observed the changes in the distance between these two stars and in the direction of the line joining them, during the year. Fig. 76 will serve to explain the general principle of this method. Let S be the position of the star whose parallax is sought, and s the position of the smaller star, both being projected on the surface of the celestial sphere. By the motion of the earth in its orbit the star S will describe the parallactic ellipse $ADBC$, and the star s, the ellipse $adbc$. When S appears to be at A, s will appear to be at a; when S is at D, s will be at d, &c. It is evident that Aa and Bb will lie

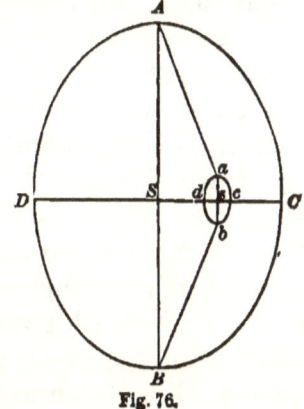

Fig. 76.

in different directions, and that Dd will be greater than Cc: and, therefore, by observing the different directions of the line joining the two stars, and also its different values, during the year, we may obtain the difference of parallax of the two stars, and, approximately, the parallax of S.

277. *Results.*—This method was applied by Bessel to the star 61 Cygni. For the sake of greater accuracy he made use of two very small stars, situated very near to that star, whose absolute parallaxes he assumed to be equal. The parallax which he obtained for this star was $0''.35$. Other observations make it $0''.56$: and the mean of these two values, or $0''.45$, corresponds to a distance of about forty-two trillions of miles, a distance which light would require seven years to traverse. The parallaxes of eleven other stars have been obtained with more or less accuracy. The star about whose parallax there is the least doubt is a Centauri, which is probably the nearest to us of all the stars. Its parallax is $0''.92$: corresponding to a distance of about twenty-one trillions of miles. A table of these stars is given at the end of this book.

According to the Russian astronomer Peters, the mean parallax of the stars of the first magnitude is $0''.21$, corresponding to a distance which light would traverse in $15\frac{1}{2}$ years. Another Russian astronomer, Struve, concludes that the distance of the most remote stars which can be seen in Lord Rosse's great telescope is about 420 times the distance of the stars of the first magnitude: from which the marvellous inference is drawn that the distance of the most remote telescopic stars from the earth is only traversed by light in 6500 years.

A knowledge of the distance of a star, and of its proper motion, enables us to estimate the amount in miles of its transverse motion. Sirius, for instance, has a proper motion of $1''.25$ a year, and its parallax is $0''.23$. Hence its annual transverse motion is equal in amount to the radius of the earth's orbit multiplied by $1\frac{25}{23}$: which is a motion of about sixteen miles a second. This is only the projection upon the celestial sphere of its real motion, which may be much greater.

278. *Real Magnitudes of the Stars.*—Hitherto, when we have determined the distance of a celestial body, we have been able to

compute its real diameter by means of observations made upon its angular diameter. But this method fails when we attempt to make use of it in obtaining the magnitude of the stars, since they do not present any measurable disc. It is true that with the better class of telescopes some of the stars appear to have a sensible disc; but this disc is really what is called a *spurious* one. This is proved by the fact that when a star is occulted by the moon, the size and the shape of the apparent disc remain unaltered up to the time of occultation, and its disappearance is then instantaneous. In the case of a solar eclipse, however, or of the occultation of a planet, the disappearance of the disc is gradual.

We can, however, obtain some idea of the probable magnitude of a star by comparing the light which it emits with that which is emitted by the sun. This comparison is made by means of the light of the moon. The ratio of the light of the sun to that of the full moon, and the ratio of the light of the full moon to that of some of the stars, have been obtained by appropriate experiments; and, from a comparison of the results of these experiments, it is inferred that if the sun were removed to a distance from the earth equal to that of the nearest fixed star, it would appear only as a star of the second magnitude. The probability, then, is that unless there is a marked difference in the intensity of the light which these different bodies emit, the sun is not so large as most, and perhaps all, of the stars of the first magnitude. There is, however, as might be expected from the delicacy of the observations, some discrepancy in the results of these various photometric experiments: the light of Sirius, for instance, is said by some observers to be one hundred, and by others to be four hundred, times as great as the light of our sun would be, were it removed to a distance from us equal to that of Sirius.

VARIABLE AND TEMPORARY STARS.

279. *Variable Stars.*—There are certain stars which exhibit periodic changes in their brightness, the periods being in some cases only a few days in length, and in other cases embracing many years. The star o *Ceti*, called also *Mira*, is an example of this class of stars. When brightest, it is a star of the second

magnitude. It remains in this state for about two weeks, and then begins to diminish in brightness, becoming wholly invisible to the naked eye in about three months, and appearing in telescopes as a star of the ninth or the tenth magnitude. After about five months it again appears, and in three months again reaches its maximum of brightness. The period in which these changes occur is 331⅓ days: at least, that is its mean value: its extreme values being 25 days more and 25 days less than this. It is also noticed that the rate of its increase and decrease of brightness is not always the same, and that one maximum of brightness is not always equal to another.

Algol, or *β Persei*, is another remarkable variable star. It remains for about sixty-one hours as a star of the second magnitude. At the end of that time it begins to decrease in brightness, and becomes a star of the fourth magnitude in less than four hours. After about twenty minutes its brightness begins to increase, and another period of less than four hours brings it up again to a star of the second magnitude.

The whole number of variable stars is between 150 and 200.

280. Several theories have been advanced in explanation of this periodicity of brightness in the variable stars. One theory is that the surfaces of these stars are not uniformly luminous, and that therefore, in rotating upon their axes, they may present at one time the lighter portions of their surfaces to the earth, and the darker portions at another. Such a variation of luminosity might be caused by the presence of spots on the surfaces of the stars, similar to the spots on the sun, but of much greater extent. A second theory is, that nebulous bodies may revolve as satellites about these stars, and may intercept their light by coming in between them and the earth. The irregularities noticed in the periods of these stars, however, seem to constitute an objection to this second theory, while they may be allowed by the first theory, since analogous fluctuations in the periods and the magnitudes of the sun's spots have been observed. Arago suggests that, if it is true, as has been asserted by some astronomers, that these stars when at their minimum are surrounded by a kind of fog, the diminution of light may be due to the interference of clouds.

281. *Temporary Stars.*—There are stars which have appeared

at times in different parts of the heavens, and have afterwards disappeared. Such stars are called *temporary* stars. In comparing recent catalogues of stars with the catalogues of ancient astronomers, it is found that some stars which were formerly visible are no longer to be seen, while others which are now visible to the naked eye are not mentioned in the ancient catalogues. Some of these cases may be due to errors of observation, but hardly all of them. Moreover, similar instances have occurred in modern times. For example, it is recorded by the Danish astronomer, Tycho Brahé, that in November, 1572, a brilliant star suddenly blazed forth near the constellation Cassiopeia, and remained in sight about 17 months. When at its brightest phase, it equalled Venus in splendor, and was visible in broad daylight. It disappeared in 1574, and has never since been seen. Similar temporary stars are recorded as having appeared in or near the same place, in 945 and 1264. It is therefore possible that this may be really a variable star, and that it may reappear in the latter part of the present century.

In May, 1866, a star of the ninth magnitude, in the constellation Corona Borealis, suddenly increased in brightness, and then rapidly decreased. On the 12th of the month it was of the second magnitude; on the 14th, of the third; and the rate of decrease in its brightness was for some time about half a magnitude a day. Lockyer says that there is good reason to believe that this sudden increase of brilliancy was due to the ignition of hydrogen in the star's atmosphere.

It is very probable that these temporary stars are really in no respect different from variable stars, except in the length of their periods. Sir John Herschel says, with reference to these stars, that it is worthy of notice that all of them which are on record have been situated in or near the borders of the Milky Way.

DOUBLE AND BINARY STARS.

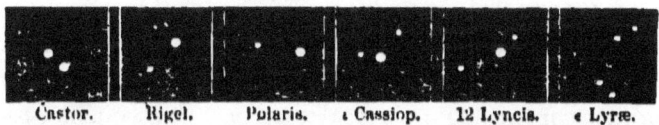

Castor. Rigel. Polaris. ι Cassiop. 12 Lyncis. ε Lyræ.

282. *Double Stars*.—Many of the stars which appear single to the naked eye, are found, when examined in telescopes, to consist of two stars, apparently very near to each other. These are called *double* stars. Only four were known to exist until near the close of the last century, when Sir William Herschel discovered about 500. The whole number of double stars now known exceeds 9000. In some cases the two stars are nearly of the same magnitude, but more frequently one of them is a large star, and the other a small one. Castor is an instance of the former class, and Sirius, Vega, and the pole-star are instances of the latter class. Some stars are found to consist of three, four, five, or more stars, and are called *triple, quadruple,* &c. stars. The star ε Lyræ, for instance, appears in ordinary telescopes to consist of two stars, but with telescopes of greater power each of these stars is resolved into two others.

283. *Binary Stars.*—The question arises with reference to these combinations: are the stars which compose them really connected, as are the sun and the planets; or is their appearance merely an optical illusion, arising from the fact that the stars in any one combination happen to lie in the same direction from the earth, although they may at the same time be at an immense distance from each other? The chances, at all events, are very much against the latter supposition. The astronomer Struve has calculated that there is only about one chance in 9570 that, if the stars of the first seven magnitudes were scattered at random in the heavens, any two of them would fall within 4" of each other; and yet more than 100 such cases have been observed. He has further calculated that there is only about one chance in 200,000 that three stars would accidentally fall within 30" of each other, so as to form a triple star; and at least four such cases are to be found.

The chances, then, are that the stars in these various combinations are physically connected. But more than this: it was announced by Sir William Herschel in 1803, after twenty-five years of observation upon many of the double stars, that in each double star which he had examined, the two stars of which it was composed revolved about each other in regular orbits, and in fact constituted a sidereal system. Subsequent observations by other astronomers have fully verified this conclusion, and about 600 double stars have been found to consist of stars revolving about each other, or rather about their common centre of gravity, according to the Newtonian law of gravitation. Such double stars are called *binary* stars, to distinguish them from other double stars, the components of which have not as yet been found to be physically connected. There are also other double stars, the components of which, while as yet they do not seem to revolve about each other, have constantly the same proper motion: thus showing that they are in all probability moving as one system through space. Triple, quadruple, &c. stars, whose constituents are found to be physically connected, are called *ternary, quaternary*, &c. stars.

The orbits of the binary stars are found to be ellipses of considerable eccentricity. The periods of their revolutions have also been approximately determined, and extend over a very wide range. There are only about twelve stars whose periods are less than 100 years, and only about 150 whose periods are less than 1000 years.

284. *Alpha Centauri.*—The star α Centauri, a star of the first magnitude in the southern hemisphere, is found to consist of two components, one of the first magnitude and the other of the second. The relative positions of these two components have been carefully noted during the last 40 years. In Fig 77, A represents one of those components, and $B\ B'\ B''$ the apparent path of the other about it. The major axis of the orbit is about 40″, and the period 75 years. Its eccentricity is 0.63.

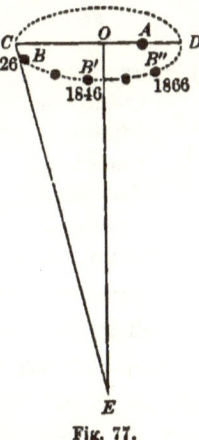

Fig. 77.

Since the distance of α Centauri from the earth is approximately known, we can obtain some idea of the dimensions of the orbit. If E in the figure represents the earth, we shall have, in the right-angled triangle COE, the angle E equal to 20″, and the side EO equal to 21 trillions of miles. Hence,

$$CO = OE \times \tan 20'' = 2{,}000{,}000{,}000 \text{ miles}:$$

which is about equal to the distance of Uranus from the sun, or to 22 times the radius of the earth's orbit.

Again, knowing the radius of the orbit and the period, we can obtain an approximate value of the mass of α Centauri from the formula in Art. 214. It will be found to be about twice the mass of the sun.

COLORED STARS.

285. Many of the stars, both isolated and double, shine with a colored light. The isolated colored stars are usually red or orange: blue or green stars being very uncommon. The number of red stars now recognized is at least 300, about 40 of which are visible to the naked eye. Among the most conspicious of the red stars are Aldebaran and Antares; and it is worthy of notice that nearly all the variable stars are of this color. According to Mr. Ennis's observations, Capella, Rigel, Procyon, and Spica are blue; Sirius, Vega, and Altair are green; and Arcturus is yellow.

The components of the double stars are often of different colors; blue and yellow, or green and yellow; and, less frequently, white and purple, or white and red. The following table contains a few of the many instances of such stars which might be given:—

Name.	Magnitude of Components.	Color of the Larger.	Color of the Smaller.
η Cassiopeiæ	4.........7	Yellow.	Purple.
α Piscium	5.........6	Pale Green.	Blue.
ι Cancri	5.........8	Orange.	Blue.
ε Boötis	3.........7	Pale Orange.	Sea Green.
ζ Coronæ	5.........6	White.	Purple.
β Cygni	3.........7	Yellow.	Blue.

When the colors of the components are complementary, and the components are of very unequal size, it is possible that only one of the colors may really exist; the other being, according to a law of Optics, merely the result of contrast. Such an opinion is held by some astronomers; but the objection is raised to it by others that, if one of these colors is only accidental, it ought to disappear when the eye is shielded from the light of the star

UNIV. OF
CALIFORNIA

PLATE IV.

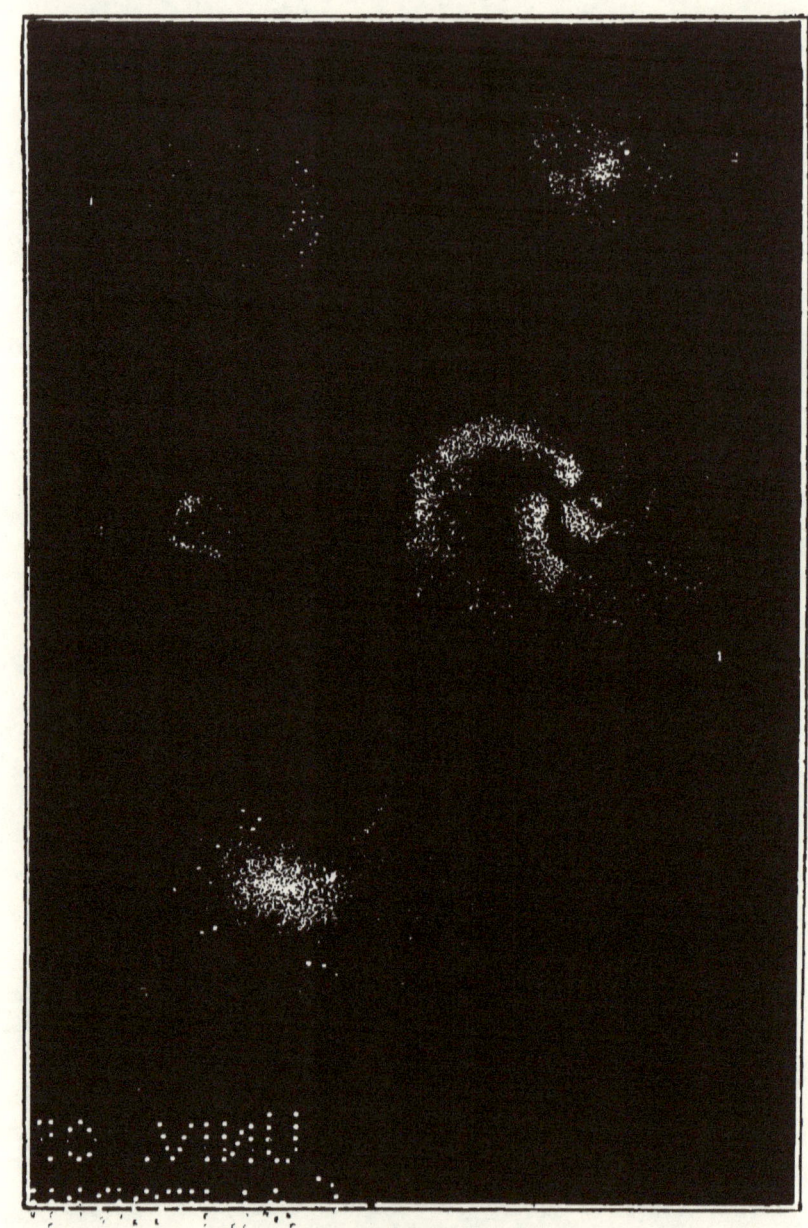

NEBULÆ.
1. ANNULAR NEBULA, 57 M LYRÆ.
2. PLANETARY NEBULA, 3614 H VIRGINIS.
3. NEBULOUS STAR, ι ORIONIS.
4. SPIRAL NEBULA, 99 M VIRGINIS.
5. CRAB NEBULA IN TAURUS.

which has the other color: which, however, is very far from being the case. Another objection is that a similar phenomenon ought to be seen in all colored stars whose components are of unequal sizes: whereas many double stars are found in which both components have the same color, sometimes red and sometimes blue. In one of Struve's catalogues, out of 596 double stars, there were:—

  295 pairs, both white:
  118 pairs, both yellowish or reddish:
  63 pairs, both bluish:
  120 pairs of totally different colors.

In a few instances stars have changed their color. Sirius, for instance, is now green; but both Ptolemy and Seneca expressly state that in their day it had a reddish hue. Capella, which is now blue, was formerly red. Some observers state that seventeen stars of the first magnitude are colored, and that seven of these have changed their color. This change of color is particularly noticeable in the variable stars. In that of 1572 (Art. 281), the color changed from white to yellow, and then to a decided red; and it is observed generally in the variable stars that the redness of the light increases as its intensity diminishes.

CLUSTERS AND NEBULÆ.

286. If we examine the heavens on any clear night when the moon is below the horizon, we shall find here and there groups of stars, which present a hazy, cloud-like appearance. These groups are classified into *clusters* and *nebulæ*, although it is difficult to establish any precise distinction between the two classes. When the different members of a group, or at all events some of them, can be separated from each other with the naked eye, the group is called a *cluster*. A *nebula*, on the other hand, is either wholly invisible to the naked eye, or, if seen at all, presents, as the name implies, only an ill-defined, cloudy appearance. The number of nebulæ which have been discovered is over 5000. Some of these nebulæ are resolved, with the aid of the telescope, into separate stars: while others, even when examined in the most powerful telescopes, preserve their cloudy appearance, and

give no trace of stars. The former are called *resolvable* nebulæ, the latter *irresolvable* nebulæ.

As every increase of telescopic power has rendered resolvable some of the nebulæ previously classified as irresolvable, the common opinion of astronomers, until within the last few years, was that there was no distinction between these nebulæ more fundamental than that of distance, magnitude, or intensity of light. Recently, however, the light of some of these nebulæ has been subjected to spectroscopic analysis, and the results seem to show the existence of decidedly different classes of nebulæ. The researches of Mr. Huggins show that the light of some of these nebulæ gives spectra, which are distinguished, not by dark lines, but by bright ones: which shows, as we have already noticed (Art. 102), that the light does not come from solid bodies in a state of ignition, but from incandescent vapor or gas. Some at least, then, of the nebulæ are to be considered as bodies of a gaseous nature; and there are indications that the gases of which they are composed are hydrogen and nitrogen. The Great Nebula in Orion is an instance of this class of nebulæ.

The nebulæ are very unequally distributed over the heavens. They congregate especially in a zone crossing the Milky Way at right angles, and they are especially abundant in the constellations Leo, Ursa Major, and Virgo.

287. *Examples of Clusters.*—The most noticeable cluster is the well-known group called the Pleiades. This group contains six or seven stars which are visible to the naked eye, the brightest star being of the third magnitude: and glimpses of many more can be obtained by examining the group with the eye turned sideways. With the aid of a telescope between one hundred and two hundred stars can be detected.

The Hyades are a group in Taurus, near the star Aldebaran, in which from five to seven stars can be counted. This group was supposed by the ancients to have some influence upon the rain.

Præsepe, or the Bee-hive, in Cancer, is visible to the naked eye as a luminous spot. With a telescope of moderate power, more than forty conspicuous stars are seen within a space about 30' square, together with many smaller ones. Before the invention of the telescope, this group was probably one of the few recognized nebulæ.

288. *Different Forms of Nebulæ.*—The groups which usually go by the name of nebulæ present, as might be expected, almost every variety of form; and most of these forms are too irregular to admit of any classification or description. Such is not the case, however, with all of them: and some of the different varieties of shape are exemplified in the following list:—

    1. Annular Nebulæ;
    2. Elliptic Nebulæ;
    3. Spiral Nebulæ;
    4. Planetary Nebulæ;
    5. Nebulous stars.

There are also individual nebulæ whose names indicate the general appearance which they present: such are the "Horse-shoe Nebula," the "Crab Nebula," the "Dumb-bell Nebula," &c.

289. *Annular Nebulæ.*—Of annular or ring-shaped nebulæ, the heavens present four examples. The most remarkable one is situated in the northern constellation Lyra. Sir John Herschel says of it: "It is small, and particularly well-defined, so as to have more the appearance of a flat, oval, solid ring than of a nebula. The axes of the ellipse are to each other in the proportion of about four to five, and the opening occupies rather more than half the diameter. The central vacuity is filled in with faint nebulæ, like a gauze stretched over a hoop. The powerful telescopes of Lord Rosse resolve this object into excessively minute stars, and show filaments of stars adhering to its edges."

290. *Elliptic Nebulæ.*—There are several instances of elliptic or oval nebulæ, of various degrees of eccentricity. The very conspicuous nebula called "the Great Nebula in Andromeda" is an example of this class. This object is distinctly visible to the naked eye, and is sometimes mistaken for a comet. When viewed with a telescope of moderate power, it has an elongated oval form; but in the largest telescopes its aspect is very different from this. Professor G. P. Bond traced it through a length of 4°, and a breadth of 2½°, and also discovered in it two curious black streaks, lying nearly parallel to the major axis of the oval. He also succeeded in detecting in it evidence of a stellar constitution.

Some elliptic nebulæ are remarkable for having double stars at or near each of their foci.

291. *Spiral Nebulæ.*—Observations with the Earl of Rosse's telescope have led to the discovery of nebulæ which consist of spiral bands proceeding from a common nucleus, and sometimes from two nuclei. The most remarkable of these spiral nebulæ is situated near the extremity of the tail of the Great Bear. It consists of nebulous coils diverging from two luminous centres, about 5' apart, and gives evidence of stellar composition.

292. *Planetary Nebulæ.*—Planetary nebulæ received their name from Sir William Herschel, on account of the resemblance in form which they bear to the larger planets of our system. The outlines of some of them are well defined, while those of others are indistinct; and some of them shine with a blue light. About twenty-five have been discovered, most of them being situated in the southern hemisphere. One of these nebulæ is situated near β Ursæ Majoris, and has a diameter of 2' 40''. It was discovered by Méchain, in 1781, and is described as "a very singular object, circular and uniform, which after a long inspection looks like a condensed mass of attenuated light." Perforations and a spiral tendency have been detected in it by the Earl of Rosse.

These planetary nebulæ can hardly be globular clusters of stars, as they would in that case be brighter in the middle than at the borders, instead of being, as they are, uniformly bright throughout. Some astronomers suppose that they are assemblages of stars, arranged either in cylindrical beds, or in the form of hollow spherical shells.

293. *Nebulous Stars.*—Nebulous stars are stars which are surrounded by a faint nebulous envelope, usually circular in form, and sometimes several minutes in diameter. In some cases this envelope terminates in a distinct outline, while in others it gradually fades away into darkness. According to Hind, nebulous stars "have nothing in their appearance to distinguish them from others entirely destitute of such appendages; nor does the nebulous matter in which they are situated offer the slightest indications of resolvability into stars, with any telescopes hitherto constructed."

294. *Double Nebulæ.*—M. D'Arrest, of Copenhagen, mentions

fifty double nebulæ whose components are not more than 5' apart, and estimates that there may be as many as 200 such double nebulæ. The two components are generally of a globular form.

It is possible that these components may in some cases be physically connected. In one instance there seem to have been changes in the distance and the relative position of the two members during the last eighty years, which indicate the possibility of a revolution of one of the members about the other.

295. *Magellanic Clouds.*—These two nebulæ, called also *nubecula major* and *nubecula minor*, are situated near the southern pole of the heavens. They are visible to the naked eye at night, when the moon is not shining, and have an oval shape. One of them covers an area of forty-two square degrees, the other an area of ten square degrees. Sir John Herschel found them to consist of swarms of stars, clusters, and nebulæ of every description.

296. *Variation of Brightness in the Nebulæ.*—In the case of a few of the nebulæ a change of brightness has been discovered, which is to some extent analogous to what has already been noticed in connection with the variable and the temporary stars. In 1852, Mr. Hind discovered a small nebula in the constellation Taurus. This nebula was observed from that time until the year 1856; but in 1861 it had entirely disappeared. In 1862 it was seen in the telescope at Pulkowa, and for a short time appeared to be increasing in brightness; but towards the end of 1863, Mr. Hind, upon searching for it with the telescope with which it was originally discovered, was unable to find any trace of it. It is a curious coincidence that a star situated very near to this nebula has changed, in the same interval of time, from the tenth magnitude to the twelfth.

There are four or five instances on record of similar changes in the brightness of nebulæ. The nebula known as 80 of Messier's Catalogue, is said to have changed, in the months of May and June, 1860, from a nebula to a well-defined star of the seventh magnitude, and then to have resumed its original appearance. The change was, however, so rapid and so decided, that some astronomers are inclined to ascribe it to the existence of a variable star between the nebula and the earth, rather than to a variation in the nebula itself. The Great Nebula in the constellation Argo, when observed by Sir John Herschel in 1838, contained in its densest part the star η, which was then of the first magnitude. Observations made in 1863, however, showed that this star had become entirely free from its nebulous envelope, and had also been reduced to a star of the sixth magnitude. The outline of certain parts of the nebula was also found to be different from what it had been represented to be by Herschel. More recent observations show still further changes in this nebula.

297. *The Galaxy, or Milky Way.*—The Milky Way, that well-known luminous band which stretches through the heavens, may be considered to be a continuous resolvable nebula. If we follow the line of its greatest brightness, and neglect occasional deviations, we find its course to be nearly that of a great circle of the heavens, inclined to the equinoctial at an angle of about 63°. At one point a kind of branch is sent off, which unites again with the main stream at a distance of about 150° from the point of separation. The angular breadth of the belt varies from 6° to 20°, the average breadth being about 10°. When examined with the telescope, this band is found to be made up of thousands and millions of telescopic stars, grouped together with every degree of irregularity. Of the 20,000,000 stars of the first fourteen magnitudes, about nine-tenths are in or near the Milky Way. At some points the stars are so close to each other that they form one bright mass of light; while at other points there are dark spaces containing scarcely a star which is visible to the naked eye. A very noticeable break of this kind occurs in that part of the Milky Way which lies nearest to the south pole. This dark space is about 8° in length and 5° in breadth. It contains only one star visible to the naked eye, and that is a very small one. Early southern navigators gave to this vacancy the name of *the Coal Sack*.

298. *Number of the Stars.*—Scarcely more can be said of the probable number of the stars, than that it is as inconceivably great as are the distances by which they are separated from the earth and from each other. Sir William Herschel states that on one occasion, while observing the stars in the Milky Way, he

estimated that 116,000 stars passed through the field of his telescope in fifteen minutes; and that, on another occasion, nearly 260,000 stars passed in forty-one minutes. It may assist us in correctly appreciating the magnitude of these numbers to remember that the number of stars visible to the naked eye in the whole heavens is less than 7000 (Art. 266). Results still more astonishing follow from an examination of the nebulæ. Take, for instance, the "Great Nebula in Andromeda," which, by observations made with the telescope at Cambridge, Massachusetts, within the last few years, has been found to give evidence of stellar composition. According to Professor G. P. Bond, its length is 4°, and its breadth $2\frac{1}{2}$°. Now, if a telescope just suffices to resolve a nebula into separate stars, it is safe to assume that the angular distance between any two contiguous stars in it cannot be greater than 1". If, then, we multiply together the number of seconds in the length of this nebula, and the number of seconds in its breadth, we obtain over 129,000,000 for an approximate estimate of the number of stars in this one nebula.

MOTION OF THE SOLAR SYSTEM IN SPACE.

299. We have now accustomed ourselves to consider the earth to be a planet, rotating upon a fixed axis once in every sidereal day, and revolving about the sun once in every sidereal year; but we have up to this point considered the sun and the solar system to be at rest in space. There are, however, observations which tend to show that such is not, in truth, the case, but that the whole solar system is in rapid motion through space. Analogy certainly supports such a conclusion. We have already seen that the solar system, immense as it may seem to be in itself, must be considered as scarcely more than a point in comparison with the entire celestial system (Art. 275); and we have also seen reasons for concluding (Art. 278) that there are many bodies among the stars which are very much larger than the sun. There is, then, no reason for supposing that the solar system is the centre of the celestial system; and there is nothing violent in the conclusion that, just as Jupiter revolves about the sun, and carries its satellites with it, so the sun may

revolve about some other body or point, and carry its system of planets and satellites with it.

300. *The Apparent Motions of the Stars.*—The proper motions of the stars, already mentioned, may be explained in three distinct ways. First, they may be what they seem to be, real motions of the stars, performed in orbits of immense radii; or, secondly, they may be only apparent motions, caused by a change of the position of the solar system in space; or, thirdly, they may be the results of both these causes existing together. The last of these theories was advanced by Sir William Herschel, in 1783; and although doubts were afterwards expressed as to its truth, later observations have tended to support it, and it is now generally accepted by astronomers. If we suppose for a moment that the solar system is approaching any given point in the heavens, the pole-star, for instance, the result will be that the stars which lie about that point will appear to separate slowly from it, while the stars which lie about the diametrically

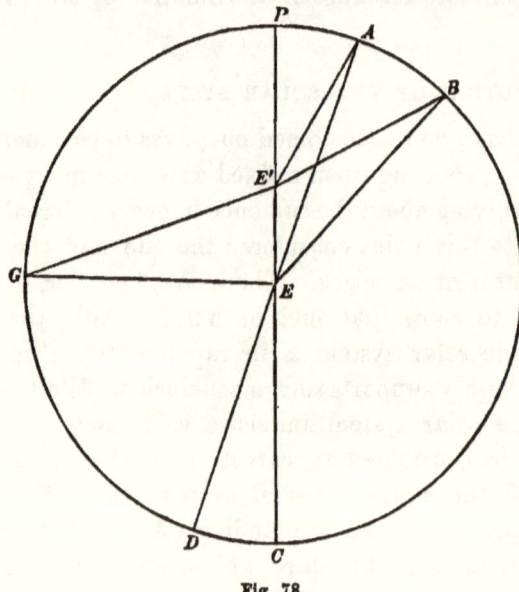

Fig. 78.

opposite point will appear to close up around that point; and indeed all the stars will apparently move in an opposite direction to that in which we suppose the solar system to be moving, those stars having the greatest motion which lie in a direction at right angles to the direction of the motion of the solar system. In Fig. 78, let E be the position in space of the earth at any time, and let $P, A, B, D,$ and G be stars supposed to be situated at equal distances from the earth. Let the earth, by the motion of the whole solar system in space, be carried to some point E', in the direction of the star P. It is evident that the angular distance of the star A from P, when the earth is at E', or the angle $PE'A$, is greater than the angular distance between A and P when the earth is at E, or the angle PEA. In other words, while the earth has moved from E to E', the star A has apparently receded from P. So also the star D has apparently approached the star C. The stars B and G have both receded from P; the retrogradation of G, or the angle EGE', being greater than the retrogradation of B, or the angle EBE'.

301. *Direction and Amount of this Motion.*—The elements, then, of the motion of the solar system through space are determined from what are called the proper motions of the stars. Recent observations and calculations have not only confirmed Herschel's theory that such a motion really exists, but have also very nearly confirmed his conclusion as to the point towards which the motion is directed. Different astronomers give different positions to this point; but they all agree in the general conclusion, that the solar system is at present moving in the direction of a point in the constellation Hercules, situated in about $17\frac{1}{3}$ hours of right ascension, and 35° of north declination.

The calculations of M. Otto Struve lead him to the conclusion that for a star situated at right angles to the direction of the motion of the solar system and at the mean distance of the stars of the first magnitude, the annual angular displacement of the star due to that motion is 0."34: that is to say, the distance through which the system moves in one year subtends at the star an angle of that amount. Now, the mean parallax of the stars of the first magnitude, or the angle subtended at the mean distance of those stars by the radius of the earth's orbit, is estimated to be 0."21 (Art. 277); hence the annual motion of the solar system is the radius of the earth's orbit multiplied by $\frac{34}{21}$, or about 150,000,000 miles: a little less than five miles a second.

This is the estimate commonly accepted by astronomers. Mr. Airy, however, deduces a motion of about twenty-seven miles a second.

302. *The Orbit of the Solar System.*—Although the motion of the solar system through space appears to be rectilinear, it does not follow that such is actually the case: since it may move in an elliptical orbit, with a radius vector so immense that years and even centuries of observation will be needed to show any sensible change of direction. The probability is that the sun, with its planets and their satellites, revolves about the common centre of gravity of the group of stars of which it is a member; and that this centre of gravity is situated in or near the plane of the Milky Way. If we suppose the orbit in which our system is moving to be an ellipse with a small eccentricity, then the centre of this ellipse will lie in a direction which makes an angle of about 90° with the direction in which the system is now moving. Mädler concluded that Alcyone, the brightest star in the Pleiades, was the central sun of the celestial sphere, while Argelander, believing that this central point must lie in the principal plane of the Milky Way, places it in the constellation Perseus. It may be noticed, in connection with this subject, that Sir William Herschel was inclined to believe that some of the more conspicuous of the stars, such as Sirius, Arcturus, Capella, Vega, and others, were in a great degree out of the reach of the attractive power of the other stars, and were probably, like the sun, centres of systems.

303. Years, and, in all probability, ages of observation will be needed to determine the elements of this most stupendous orbit. Mädler's speculations have led him to the conclusion that the period of this revolution is 29,000,000 years: a period in comparison with which the years through which astronomical observations have extended are utterly insignificant. The student who is curious to know more of this subject will find the details fully set forth in Grant's *History of Physical Astronomy.* Grant himself says, in reference to the subject:—"It is manifest that all such speculations are far in advance of practical astronomy, and therefore they must be regarded as premature, however probable the supposition on which they are based, or how-

ever skilfully they may be connected with the actual observation of astronomers."

DETERMINATION OF THE REAL MOTIONS OF THE STARS.

304. We have already seen (Art. 265), that in order to determine the real motion of a star in space, we must be able to determine, not only its transverse motion, which is indicated by a change in its apparent position upon the celestial sphere, but also its motion directly to or directly from the earth. Now, a star situated at the nearest distance of the fixed stars, and moving towards the earth with a velocity equal to that of the earth in its orbit (18 miles a second), would diminish its distance from us by only about $\frac{1}{30}$th in a thousand years. The detection of any such motion by a change in the star's apparent brightness is, therefore, utterly out of the question. The spectroscope, however, gives us quite another means of detecting such a motion.

305. *Analogy Between Light and Sound.*—A clear conception of the principle upon which this method rests may be more readily obtained if we first notice the analogy between the conduction of light and that of sound. Sound is the result of a series of waves or pulses in the air, produced by the vibrations of a sonorous body; light is the result of a similar series of pulses or waves in the luminiferous ether, caused by the vibrations of the particles of a luminous body. The more rapidly a sonorous body vibrates, the greater will be the number of pulses or waves which it communicates to the air in the unit of time, and, consequently, the higher will be the pitch of the tone produced. In the case of a luminous body, the greater the number of waves in the luminiferous ether which the vibrations of any particle cause in the unit of time, the greater will be the refrangibility of the ray produced; in the solar spectrum, for instance, the violet rays are the most refrangible of all the rays which we can see, and the number of waves in the unit of time necessary to produce them is greater than the number necessary to produce rays of any other color. The refrangibility of a ray is therefore analogous to the pitch of a tone.

306. *Change of Tone or Refrangibility.*—It is proved by ex-

periment that if the distance between a sonorous body, producing a tone of constant pitch, and the hearer, is diminished by the motion of either, with a velocity that has an appreciable ratio to that of sound, the pitch of the tone will appear to be heightened; and that if, on the contrary, the distance between the two is increased with sufficient rapidity, the pitch of the tone will appear to be lowered. So, too, in the case of light: if the luminous body and the observer approach each other, the refrangibility of the rays of light will be increased; and if they separate, it will be diminished. If, then, we can detect any change of refrangibility in the light of a heavenly body, we may conclude that the distance of that body from the earth is changing.

307. *Change of Refrangibility Detected.*—Father Secchi, in a communication to the *Comptes Rendus* in the early part of 1868, after stating the principles of this method, announced that he had subjected the light of Sirius and of other prominent stars to an examination with the spectroscope, but had detected no evidence of motion. Since then, Mr. Huggins, an English observer, who has devoted himself especially to spectroscopic investigations, has succeeded in detecting such a motion in Sirius. There is a certain dark line F in the blue of the solar spectrum, which corresponds to a bright line in the spectrum of hydrogen; and this same line also appears in the spectrum of Sirius. Now, if Sirius has no motion either towards or from the earth, the line F in its spectrum will coincide in position with the corresponding line in the spectrum of hydrogen, when the two spectra are compared by means of the spectroscope: while if, on the contrary, Sirius has such a motion, its whole spectrum, lines and all, will be shifted bodily, and the line F will no longer coincide with the corresponding line in the spectrum of hydrogen.

Mr. Huggins, using a spectroscope of large dispersive power, and carefully comparing the spectrum of Sirius with that of hydrogen, finds that the line F in the spectrum of Sirius *is displaced*, by about $\frac{1}{250}$th of an inch. This displacement is towards the red end of the spectrum, and indicates that the refrangibility of the light of Sirius is diminished: since the red rays are the least refrangible of all the colored rays of the spectrum. Sirius is therefore *receding* from the earth.

308. *Amount of the Real Motion of Sirius.*—The amount of recession corresponding to a displacement of this extent, when observed in a spectroscope whose power is equal to that of the one used by Mr. Huggins, is computed to be about 41½ miles a second. But it happens that when the observation was made, the earth was so situated in its orbit that it was receding from Sirius, by its revolution in its orbit, at the rate of about 12 miles a second. The motion of the solar system in space, computed to be five miles a second (Art. 301), must also be taken into consideration; and the point towards which this motion is at present directed is almost exactly opposite to the position of Sirius on the celestial sphere. The earth was therefore moving away from Sirius at the rate of about 17 miles a second. If, then, we diminish the whole amount of the increase of distance between Sirius and the earth in one second, by the amount of the motion of the earth through space in one second, or 17 miles, we find that Sirius is moving through space, away from the earth, at the rate of about 24½ miles a second.

By combining this motion with the transverse motion of Sirius, we can obtain an approximate value of its real motion. In Art. 277, the transverse motion of Sirius was computed to be about 16 miles a second. The resultant of these two motions is a motion of about 29 miles a second: or, 900,000,000 miles a year.

Mr. Huggins further concludes that five stars of the Dipper are receding from us, and that the bright star Arcturus is approaching us.

309. The numerical results of the preceding article may not be beyond criticism; but the grand fact remains, that in all probability the motions of these distant bodies, which have so long seemed to be wrapped in hopeless mystery, are soon to come within the reach of our observation. The knowledge of these motions will be a powerful auxiliary in the determination of the law which undoubtedly binds all the heavenly bodies together in one great system; and it is not presumptuous to expect that at some future day—no one can say how distant or how near—this law will be revealed to us.

APPENDIX.

MATHEMATICAL DEFINITIONS AND FORMULÆ.

PLANE TRIGONOMETRY.

1. The *complement* of an angle or arc is the remainder obtained by subtracting the angle or arc from 90°.

2. The *supplement* of an angle or arc is the remainder obtained by subtracting the angle or arc from 180°.

3. The *reciprocal* of a quantity is the quotient arising from dividing 1 by that quantity: thus the reciprocal of a is $\dfrac{1}{a}$.

4. In the series of right triangles ABC, $AB'C'$, $AB''C''$, &c., Fig. 79, having a common angle A, we have by Geometry,

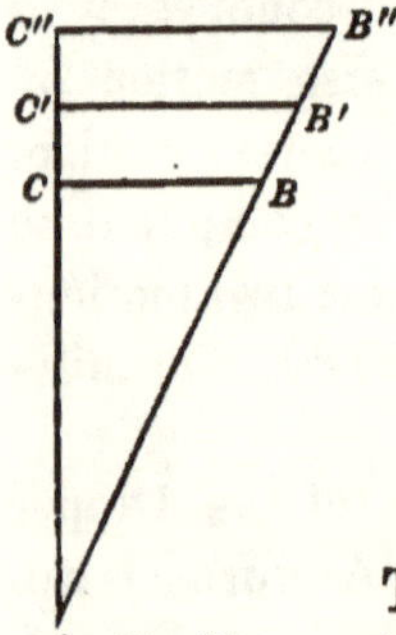

$$\frac{BC}{AB} = \frac{B'C'}{AB'} = \frac{B''C''}{AB''}.$$

$$\frac{BC}{AC} = \frac{B'C'}{AC'} = \frac{B''C''}{AC''}.$$

$$\frac{AB}{AC} = \frac{AB'}{AC'} = \frac{AB''}{AC''}.$$

Fig. 79.

The ratios of the sides to each other are therefore the same in all right triangles having the same acute angle: so that, if these ratios are known in any one of these triangles, they will be known in all of them. These ratios, being thus dependent only on the value of the angle, without any regard to the absolute lengths of the sides, have received special names, as follows:

The *sine* of the angle is the quotient of the opposite side divided by the hypothenuse. Thus, $\sin A = \dfrac{BC}{AB}$.

The *tangent* of the angle is the quotient of the opposite side divided by the adjacent side. Thus, $\tan A = \dfrac{BC}{AC}$.

The *secant* of the angle is the quotient of the hypothenuse divided by the adjacent side. Thus, $sec\ A = \dfrac{AB}{AC}$.

5. The *cosine, cotangent,* and *cosecant* of the angle are respectively the sine, tangent, and secant of the complement of the angle. Now, in Fig. 79, the angle ABC is evidently the complement of the angle BAC. Hence we have,

$$cos\ A = sin\ B = \dfrac{AC}{AB};$$

$$cot\ A = tan\ B = \dfrac{AC}{BC};$$

$$cosec\ A = sec\ B = \dfrac{AB}{BC}.$$

6. Since the reciprocal of $\dfrac{a}{b}$ is $\dfrac{b}{a}$, we see, from the preceding definitions, that the sine and the cosecant of the same angle, the tangent and the cotangent, the cosine and the secant, are reciprocals.

7. If two parts of a plane right triangle in addition to the right angle are given, one of them being a side, the triangle can be *solved*: that is to say, the values of the remaining parts can be obtained. This solution is effected by means of the definitions above given.

8. When an angle is very small, its sine and its tangent are both very nearly equal to the arc which subtends the angle in the circle whose radius is unity. Hence, to find the sine or the tangent of a very small angle *approximately*, we have, if x is a small angle expressed in seconds,

$$sin\ x = tan\ x = x\ sin\ 1''.$$

If x is expressed in minutes,

$$sin\ x = tan\ x = x\ sin\ 1'.$$

9. If x and y are any two small angles, we have from the preceding article,

$$\dfrac{sin\ x}{sin\ y} = \dfrac{x\ sin\ 1''}{y\ sin\ 1''} = \dfrac{x}{y};$$

that is, the sines (or tangents) of very small angles are proportional to the angles themselves.

10. $Cos\ x = 1 - 2\ sin^2 \tfrac{1}{2}\ x.$

244   APPENDIX.

11. The sides of a plane triangle are proportional to the sines of their opposite angles.

12. In order to solve a plane oblique triangle, three parts must be given, and one of them must be a side.

SPHERICAL TRIGONOMETRY.

13. A *spherical triangle* is a triangle on the surface of a sphere, formed by the arcs of three great circles of the sphere.

14. In a spherical right triangle, the *sine* of either oblique angle is equal to the quotient of the sine of the opposite side, divided by the sine of the hypothenuse. Thus, in the triangle ABC, right-angled at C, Fig. 80, we have,

$$sin\ A = \frac{sin\ BC}{sin\ AB}.$$

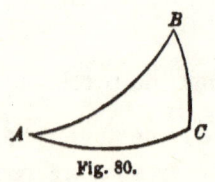
Fig. 80.

15. The *cosine* of either oblique angle is equal to the quotient of the tangent of the adjacent side, divided by the tangent of the hypothenuse. Thus,

$$cos\ A = \frac{tan\ AC}{tan\ AB}.$$

16. The *tangent* of either oblique angle is equal to the quotient of the tangent of the opposite side, divided by the sine of the adjacent side. Thus,

$$tan\ A = \frac{tan\ BC}{sin\ AC}.$$

17. A spherical right triangle can be solved when any two parts in addition to the right angle are given. The solution is effected by means of the formulæ given in the preceding articles.

ANALYTIC GEOMETRY.

18. The circle, the ellipse, the hyperbola, and the parabola are often called *conic sections;* since each curve may be formed by intersecting a right cone by a plane.

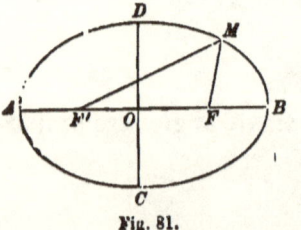
Fig. 81.

19. An *ellipse* is a plane curve, in which the sum of the distances of each point from two fixed points is equal to a given line. Thus, in Fig. 81, the

sum of the distances of the point M from the two fixed points F and F' is always constant, wherever on the curve $ACBD$ the point M may be situated.

The two fixed points are called the *foci* of the ellipse, and the middle point of the line which joins them is called the *centre*.

A line drawn from either focus to any point of the curve is called a *radius vector*.

A line drawn through the centre, and terminating at each extremity in the curve, is called a *diameter*. That diameter which passes through the foci is called the *transverse* or *major axis:* and that diameter drawn perpendicular to the transverse axis is called the *conjugate* or *minor axis*. Thus, AB is the transverse axis, and CD the conjugate axis. The transverse axis is equal to the constant sum of the distances FM and $F'M$.

The *eccentricity* of the ellipse is the quotient of the distance from the centre to either focus, divided by half the major axis. Thus, $\dfrac{OF}{OB}$ is the eccentricity.

20. The solid generated by the revolution of an ellipse about its major axis is called a *prolate spheroid:* that generated by its revolution about its minor axis is called an *oblate spheroid*.

The expression for the volume of an oblate spheroid is $\frac{4}{3}\pi a^2 b$: in which a and b denote the semi-major and the semi-minor axis of the generating ellipse, and π the ratio of the circumference of a circle to its diameter, or 3.1416.

The *compression*, or *oblateness*, of an oblate spheroid is the ratio of the difference between the major and the minor axis of the generating ellipse to the major axis.

21. The *hyperbola* is a plane curve, in which the difference of the distances of each point from two fixed points is equal to a given line. Thus, in Fig. 82, the difference of the distances of any point M of the curve from the two fixed points F and F' is equal to a given line.

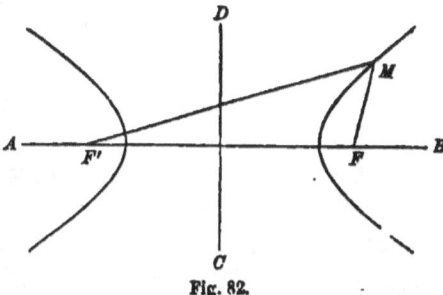

Fig. 82.

The two fixed points are called the *foci*, and the middle point of the line joining them is called the *centre*.

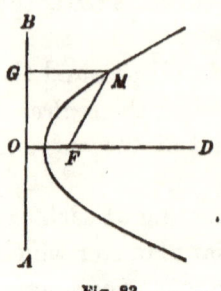

Fig. 83.

22. The *parabola* is a plane curve, every point of which is equally distant from a fixed point, and from a right line given in the plane. Thus, in Fig. 83, in which AB is the given line, and F the given point, the distances FM and GM are equal to each other, for any point M of the curve.

The fixed point is called the *focus*.

MECHANICS.

23. The *resultant* of two or more forces is a force which singly will produce the same mechanical effect as the forces themselves jointly.

The original forces are called *components*. In all statical investigations the components may be replaced by their resultant, and *vice versa*.

24. If two forces be represented in magnitude and direction by the two adjacent sides of a parallelogram, the diagonal will represent their resultant in magnitude and direction. Thus, in

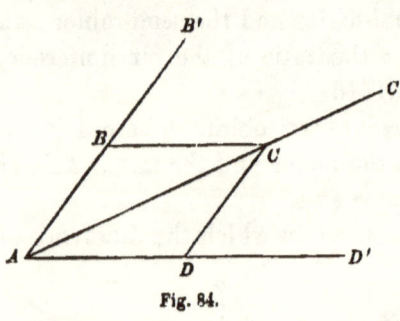

Fig. 84.

Fig. 84, if a force act at A in the direction AD', and a second force act at A in the direction AB', these two forces being represented in magnitude by the lengths of the lines AD and AB respectively, the resultant of these two forces will be a force in the direction AC', and of a magnitude represented by the length of the line AC. The parallelogram is called the *parallelogram of forces*.

25. Conversely: any given force may be resolved into two component forces, acting in given directions. Thus, in Fig. 84, let AC be the given force, and AB' and AD' the given directions. From C draw CB parallel to AD', and CD parallel to AB'.

AB and AD will be the two components acting in the given directions.

26. The force which must continually urge a material point towards the centre of a circle, in order that it may describe the circumference with a uniform velocity, is equal to the square of the linear velocity divided by the radius. This proposition is expressed in the following formula;

$$f = \frac{v^2}{r};$$

v being the space passed over in the unit of time, r the radius of the circle, and f the magnitude of the force.

27. The force which constantly urges a body towards the centre of its circular path is called a *centripetal force*. The tendency which the body has to recede from the centre, in consequence of its inertia, or the resistance which it offers to a deflection from a rectilinear path, the resistance being estimated in the direction of the radius, is called a *centrifugal force*. These two forces are in equilibrium as long as the body moves in the same circular path, and the same expression serves for the measure of each force.

28. Let t be the *periodic time*, or the time of one revolution. We shall evidently have,

$$vt = 2\pi r;$$
$$\therefore v^2 = \frac{4\pi^2 r^2}{t^2}.$$

Substituting this value of v^2 in the formula in Art. 26 above, we shall have,

$$f = \frac{4\pi^2 r}{t^2}.$$

KIRKWOOD'S LAW.

In 1848, Daniel Kirkwood, of Pennsylvania, announced the discovery of an analogy in the distances, masses, periods of rotation, and periods of revolution of the principal planets. This analogy is known under the name of *Kirkwood's Law*. The original statement of it will be found in the American Journal of Science, Vol. IX., Second Series, and is as follows:

"Let P be the point of equal attraction between any planet and the one next interior, the two being in conjunction: P' that between the same and the one next exterior.

"Let also $D =$ the sum of the distances of the points P, P' from the orbit of the planet, which I shall call the diameter of the sphere of the planet's attraction:

"$D' =$ the diameter of any other planet's sphere of attraction found in like manner:

"$n =$ the number of sidereal rotations performed by the former during one revolution around the sun:

"$n' =$ the number performed by the latter; then it will be found that

$$n^2 : n'^2 = D^3 : D'^3; \text{ or } n = n' \left(\frac{D}{D'}\right)^{\frac{3}{2}}.$$

That is, *the square of the number of rotations made by a planet during one revolution round the sun, is proportional to the cube of the diameter of its sphere of attraction:* or $\frac{n}{D^{\frac{3}{2}}}$ is a constant quantity for all the planets of the solar system."

This analogy, when first announced, created much interest among scientific men, many of whom considered that, if its truth were established, it would be a powerful argument in favor of the nebular hypothesis. There is so much uncertainty attending the determination of the masses and the periods of rotation of many of the planets, that the statement of this analogy has been excluded from the main text of this book.

ASTRONOMICAL CHRONOLOGY.

The science of Astronomy seems to have been cultivated in very early ages by almost all the Eastern nations, particularly by the Egyptians, the Persians, the Indians, and the Chinese. Records of observations made by the Chaldæans as far back as 2234 B.C. are said to have been found in Babylon. The study of Astronomy was continued by the Chaldæans, and in later times by the Greeks and the Romans, until about 200 A.D. After that time it was neglected for about six centuries, and was then resumed by the Eastern Saracens after Bagdad was built. It was brought into Europe in the thirteenth century by the Moors of Barbary and Spain. A full account of the rise and the progress of the study of Astronomy, is given in *Vince's Astronomy*, Vol. II.

The instrument principally used by the ancient astronomers seems to have been a sort of quadrant, mounted in a vertical position. Ptolemy describes one which he used in determining the obliquity of the ecliptic. The Arabian astronomers had one with a radius of $21\frac{2}{3}$ feet, and a sextant with a radius of $57\frac{3}{4}$ feet.

The following dates, with scarcely any change or addition, have been taken from George F. Chambers's *Descriptive Astronomy* (Oxford, 1867), in which many other interesting astronomical dates, here omitted, may be found. [See Note, page 254.]

B.C.
720. Occurrence of a lunar eclipse, recorded by Ptolemy. It seems to have been total at Babylon, March 19, $9\frac{1}{2}$h. P.M.

640–550. Thales, of Miletus, one of the seven wise men of Greece. He declared that the earth was round, calculated solar eclipses, discovered the solstices and the equinoxes, and recommended the division of the year into 365 days.

585. Occurrence of a solar eclipse, said to have been predicted by Thales.

580–497. Pythagoras, of Crotona, founder of the Italian Sect. He taught that the planets moved about the sun in elliptic orbits, and that the earth was round; and is said to have suspected the earth's motion.

547. Anaximander died. He asserted that the earth moved, and that the moon shone by light reflected from the sun. He introduced the sun-dial into Greece.

500. Parmenides, of Elis, taught the sphericity of the earth.

490. Alcmæon, of Crotona, asserted that the planets moved from west to east.

450. Diogenes, of Apollonia, stated that the inclination of the earth's orbit caused the change of seasons. Anaxagoras is said to have explained eclipses correctly.

B.C.

432. Meton introduced the luni-solar period of 19 years. He observed a solstice at Athens in 424.
384–322. Aristotle wrote on many physical subjects, including Astronomy.
370. Eudoxus introduced into Greece the year of 365¼ days.
330. Callippus introduced the luni-solar period of 76 years. Pytheas measured the latitude of Marseilles, and showed the connection between the moon and the tides. He also showed the connection of the inequality of the days and nights with the differences of climate.
320–300. Autolychus, author of the earliest works on Astronomy extant in Greek.
306. First sun-dial erected in Rome.
287–212. Archimedes observed solstices, and attempted to measure the sun's diameter. Aristarchus declared that the earth revolved about the sun, and rotated on its axis.
276–196. Eratosthenes measured the obliquity of the ecliptic, and found it to be 20½ degrees. He also measured a degree of the meridian with considerable exactness.
190–120. Hipparchus, called the Newton of Greece. He discovered precession: used right ascensions and declinations, and afterwards latitudes and longitudes: calculated eclipses: discovered parallax: determined the mean motions of the sun and the moon; and formed the first regular catalogue of the stars.
50. Julius Cæsar, with the Egyptian astronomer Sosigenes, undertook to reform the calendar.

A.D.

100–170. Ptolemy, of Alexandria, author of the Ptolemaic System of the Universe, in which the earth is the centre. He wrote descriptions of the heavens and the Milky Way, and formed a catalogue giving the positions of 1022 fixed stars. He appears to have been the first to notice the refraction of the atmosphere.
642. The School of Astronomy, at Alexandria, founded ten centuries previously by Ptolemy the Second, was destroyed by the Saracens under Omar.
762. Rise of Astronomy among the Eastern Saracens.

ASTRONOMICAL CHRONOLOGY.

A.D.
880. Albatani, the most distinguished astronomer between Hipparchus and Tycho Brahé. He corrected the values of precession and the obliquity of the ecliptic, formed a catalogue of the stars, and first used sines, chords, &c.

1000. Abùl-Wefa first used secants, tangents, and cotangents.

1080. The use of the cosine introduced by Geber, and also some improvements in Spherical Trigonometry.

1252. Alphonso X., King of Castile, under whose direction certain celebrated astronomical tables, called the *Alphonsine Tables*, were compiled.

1484. Waltherus used a clock with toothed wheels. (The earliest complete clock of which there is any certain record was made by a Saracen in the thirteenth century.)

1543. Publication of Copernicus's *De Revolutionibus Orbium Celestium*, setting forth the *Copernican* System of the Universe.

1581. Galileo determined the isochronism of the pendulum.

1582. Tycho Brahé commenced astronomical observations near Copenhagen.

1594. Gerard Mercator, author of *Mercator's Projection*. (The date is doubtful, and may have been as early as 1556.)

1576. Fabricius discovered the variability of *o* Ceti.

1603. Bayer's Maps of the Stars published.

1604. Kepler obtained an approximate value of the correction for refraction.

1608. Jansen and Lippersheim, of Holland, are said to have invented the refracting telescope, using a convex lens. It is, however, a disputed point as to who really invented the telescope. The use of the lens seems to have been known about fifty years before this.

1609. Kepler announced his first two laws.

1610. Galileo, using a telescope with a concave object-lens, discovered the satellites of Jupiter, the librations of the moon, the phases of Venus, and some of the nebulæ.

1611. Spots and rotation of the sun discovered by Fabricius.

1614. Napier invented logarithms.

1618. Kepler announced his third law.

1631. The first recorded transit of Mercury, observed by Gassendi. The vernier invented.

APPENDIX.

A.D.
1633. Galileo forced to deny the Copernican theory.
1639. First recorded transit of Venus, observed by Horrox and Crabtree.
1640. Gascoigne applied the micrometer to the telescope.
1646. Fontana observed the belts of Jupiter.
1654. The discovery of Saturn's rings by Huyghens. (Galileo had previously noticed something peculiar in the planet's appearance.)
1658. Huyghens made the first pendulum-clock. (The discovery is also ascribed to Galileo the younger.)
1662. Royal Society of London founded.
1663. Gregory invented the *Gregorian* reflecting telescope.
1664. Hook detected Jupiter's rotation.
1666. Foundation of the Academy of Sciences at Paris.
1669. Newton invented the *Newtonian* reflecting telescope.
1673. Flamsteed explained the equation of time.
1674. Spring pocket-watches invented by Huyghens. (Said also to have been invented, somewhat earlier, by Dr. Hooke.)
1675. Transmission of light discovered by Römer. Transit Instrument used to determine right ascensions by Römer. Greenwich Observatory founded.
1687. Newton's *Principia* published.
1690. Ellipticity of the earth theoretically determined by Huyghens.
1704. Meridian Circle used by Römer.
1711. Foundation of the Royal Observatory at Berlin.
1725. Compensation pendulum announced by Harrison.
1726. Mercurial pendulum invented by Graham.
1727. Aberration of light discovered by Bradley.
1731. Hadley's Quadrant invented.
1744. Euler's *Theoria Motuum* published, the first analytical work on the planetary motions.
1745. Nutation of the earth's axis discovered by Bradley.
1750. Wright's *Theory of the Universe* published.
1765. Harrison rewarded by Parliament for the invention of the Chronometer.
1767. British Nautical Almanac commenced.
1769. Transit of Venus successfully observed.

ASTRONOMICAL CHRONOLOGY. 253

A.D
1774. Experiments by Maskelyne to determine the earth's density.
1781. Uranus discovered by Sir William Herschel. Messier's catalogue of Nebulæ published.
1783. Herschel suspected the motion of the whole solar system.
1784. Researches of Laplace on the stability of the solar system.
1786. Publication of Herschel's catalogue of 1000 nebulæ.
1787. Herschel began to observe with his forty-feet reflector. The Trigonometrical Survey of England commenced.
1788. Publication of La Grange's *Mécanique Analytique*.
1789. The rotation of Saturn determined by Herschel, and a second catalogue of 1000 nebulæ published.
1792. Commencement of the Trigonometrical Survey of France.
1796. French Institute of Science founded.
1799. Laplace's *Mécanique Celeste* commenced.
1801-7. The minor planets Ceres, Pallas, Juno, and Vesta discovered.
1802. Publication of Herschel's third catalogue of Nebulæ.
1803. Publication of Herschel's discovery of Binary Stars.
1804. Proper motions of 300 stars published by Piazzi.
1805. Commencement of attempts to determine the parallax of the stars.
1812. Troughton's Mural Circle erected at Greenwich.
1820. Foundation of the Royal Astronomical Society of London.
1821. Observatory at Cape of Good Hope founded.
1823. Cambridge (England) Observatory founded.
1835. Airy determined the time of Jupiter's rotation.
1837. Value of the moon's equatorial parallax determined by Henderson. The East Indian arc of 21° 21' completed.
1838. Parallax of 61 Cygni determined by Bessel.
1839. Parallax of *a* Centauri determined by Henderson. Imperial Observatory at Pulkowa (Russia) founded.
1840. Harvard Observatory founded.
1842. Washington Observatory founded. Mean density of the earth determined by Baily.
1843. Periodicity of the solar spots detected by Schwabe.
1844. Taylor's catalogue of 11,015 stars. Transmission of time by electric signals commenced in the United States.

A.D.
1845. Discovery of the fifth minor planet, Astræa. (187 others have since been discovered.)
1846. Discovery of the planet Neptune.
1847. Lalande's catalogue of 47,390 stars republished by the British Association.
1851. Foucault's pendulum experiment to demonstrate the earth's rotation. Completion of the Russo-Scandinavian arc.
1854. Difference of longitude of Greenwich and Paris determined by electric signals.
1855. Commencement of the American Ephemeris.
1858. De La Rue obtained a stereoscopic view of the moon. (The first photograph of the moon was made by Dr. J. W. Draper, of New York, in 1840.)
1859. Publication of Section I. of Argelander's "Zones", containing 110,982 stars. Suspected discovery of the planet Vulcan, lying within the orbit of Mercury. Completion of the Berlin Star Charts commenced in 1830.
1861. Appearance, in June, of a comet with a tail of 105° (the longest on record). Publication of Section II. of Argelander's "Zones", containing 105,075 stars.
1862. Section III of Argelander's "Zones", containing 108,129 stars. Notes on 989 Nebulæ, by the Earl of Rosse. Bond's Memoir on Donati's Comet of 1858, published in the *Annals of the Harvard Observatory*.
1863. Announcement by several computers that the received value of the sun's parallax is too small by about $0''.3$. Spectroscopic observations of celestial objects, by Huggins and Miller.
1864. Publication of Sir John Herschel's great catalogue of 5079 nebulæ.
1866. Magnificent display of shooting-stars on the morning of November 14th.

NOTE. — The third edition of Chambers's *Descriptive Astronomy* (Oxford, 1877) contains an exceedingly interesting and elaborate summary of Astronomical Chronology.

NAVIGATION.

The earliest accounts of Navigation appear to be those of Phœnicia, 1500 B.C. Long voyages are mentioned in the earliest mythical stories; but the first considerable voyage of even probable authenticity seems to be that of the Phœnicians about Africa, 600 B.C. The Roman navy dates from 311 B.C., and that of the Greeks from a much earlier time. After the decay of these nations, commerce passed into the hands of Genoa, Venice, and the Hanse towns; from them it passed to the Portuguese and the Spanish; and from them again to the English and the Dutch.

The attractive power of the magnet has been known from time immemorial; but its property of pointing to the north was probably discovered by Roger Bacon in the thirteenth century. When first used as a compass, the needle was placed upon two bits of wood, which floated in a basin of water; and the method of suspending it dates from 1302. The variation of the compass was discovered by Columbus, in 1492. The compass-box and the hanging-compass were invented by the Rev. William Barlowe, in 1608.

Plane charts were used about 1420. The discovery that the loxodromic curve is a spiral was made by Nonius, a Portuguese, in 1537. The log is first mentioned by Bourne, in 1577. Logarithmic tables were applied to Navigation by Gunter, in 1620; and middle-latitude sailing was introduced three years afterwards. Other dates relating to the subject of Navigation are given in the Astronomical Chronology.

TABLE I.

THE PLANETS, THE SUN, AND THE MOON.

Name.	Symbol.	Inclination to the Ecliptic.	Eccentricity of Orbit.	Greatest distance from Sun.	Least distance from Sun.	Mean distance from Sun.	Mean distance from Sun in miles.
		° ′ ″					
Mercury	☿	7 00 08	0.20560	0.46669	0.30750	0.38710	35,760,000
Venus...	♀	3 23 31	0.00684	0.72826	0.71840	0.72333	66,820,000
Earth ...	⊕ or ♁		0.01679	1.01679	0.98321	1.00000	92,380,000
Mars.....	♂	1 51 05	0.09326	1.66578	1.38160	1.52369	140,800,000
Jupiter.	♃	1 18 40	0.04824	5.45378	4.95182	5.20280	480,600,000
Saturn...	♄	2 29 28	0.05560	10.07328	9.00442	9.53885	881,200,000
Uranus..	⛢ or ♅	0 46 30	0.04658	20.07612	18.28916	19.18264	1,772,000,000
Neptune	♆	1 46 59	0.00872	30.29888	29.77506	30.03697	2,775,000,000
Moon....	☾	5 08 40	0.05491				

Name.	Sidereal Period in days.	Daily Heliocentric Motion.	Synodic Period in days.	Max. Diameter from ⊕.	Min. Diameter from ⊕.	Diam. at mean distance.	Diameter in miles.
		° ′ ″		″	″	″	
Mercury	87.969	4 05 33	115.877	12.9	04.5	06.7	3,000
Venus...	224.701	1 36 08	583.921	1′ 06.3	09.7	17.1	7,600
Earth ...	365.256	59 08					7,925.6
Mars.....	686.980	31 27	779.936	30.1	04.1	11.1	4,100
Jupiter.	4,332.585	4 59	398.884	50.6	30.8	37.2	89,000
Saturn...	10,759.220	2 01	378.092	20.3	14.6	16.1	72,000
Uranus..	30,686.821	42	369.656	04.3	03.5	03.9	33,000
Neptune	60,126.722	22	367.488	02.9	02.6	02.7	37,000
Moon....	27.322		29.531	33′ 31	28′ 48	31′07	2,161.6
Sun				32 35.6	31 31	32 03.6	861,400

Name.	Volume.	Mass.	Density.	Rotation.	Inclination of Axis to Ecliptic.	App. Diam. of Sun from Planet.
				d. h. m. s.		′ ″
Mercury	0.052	$\frac{1}{3000000}$	2.02	24 05 30		82 49
Venus...	0.851	$\frac{1}{400000}$	0.90	23 21 23	73° 32′	44 19
Earth ...	1.000	$\frac{1}{327000}$	1.00	23 56 04	66 33	32 04
Mars.....	0.239	$\frac{1}{3100000}$	0.50	24 37 23	61 18	21 02
Jupiter.	1,387.431	$\frac{1}{1048}$	0.23	9 55 21	86 55	6 10
Saturn...	746.898	$\frac{1}{3500}$	0.13	10 14 24	62	3 22
Uranus..	72.359	$\frac{1}{22600}$	0.18			1 40
Neptune	98.664	$\frac{1}{19000}$	0.18			1 04
Moon....	0.020	$\frac{1}{27000000}$	0.63	27 07 43 11	88 30	
Sun......	1,284,000	1	0.25	25 07 48	82 30	

Too much confidence must not be placed in the absolute accuracy of all the elements above tabulated. The necessity of this caution will become obvious to any one who will compare the values of these elements as they are given by different authorities. The relative distances, the apparent diameters, and the periods of the planets, being matters of direct observation, are known to a great degree of accuracy; but the absolute distances and diameters, and the volumes, depending as they do upon the distance of the earth from the sun, must be considered to be only approximately known. The masses, too, of some of the planets are uncertain: and so also must be the densities, which depend upon the masses.

TABLE II.

THE EARTH.

```
Density...................................................5.67 (water being 1)
Polar Diameter............................................7899.170 miles.
Equatorial   "  ..........................................7925.648    "
Compression..............................................0.00334
Length of sidereal year.................................365d. 06h. 09m. 09.6s.
   "    " tropical   "  ..................................  " 05  48   47.8
   "    " anomalistic "..................................  " 06  13   49.3
   "    " sidereal day ................................... 23  56   04.09
Eccentricity of orbit....................................0.0167917
Inclination    "    "  in 1868...........................23° 27' 23"
Annual diminution........................................0".4645
   "     precession......................................50".24
   "     advance of line of nodes........................11".8
```

TABLE III.

THE MOON.

```
Distance from earth in earth's radii.....................60.267
Mean distance from earth.................................238,800 miles.
Greatest   "       "      "  ............................257,900   "
Least      "       "      "  ............................221,400   "
Sidereal revolution......................................27d. 07h. 43m. 11.5s.
Synodical      "  ........................................29  12   44   03
Mean daily geocentric motion.............................13° 10' 36"
Revolution of nodes......................................6793.43 days.
   "        " perigee....................................3232.58   "
Mean horizontal parallax.................................57' 03"
Greatest    "        "  .................................61  32
Least       "        "  .................................52  50
Radius in terms of earth's radius........................0.2727
Mass    "    "    "    "    mass........................ 1/81
Density "    "    "    "    density..................... 3/5
```

TABLE IV.

SATELLITES.

SATELLITES OF MARS.

Names.	Inclination of orbit to orbit of Primary.	Distance from Primary in miles.	Sid. Period. d. h. m.	Diameter in miles.	Magnitude.
I. Phobus	[They move very nearly in the plane of Mars's equator.]	5,800	0 07 39	[Not yet determined, but very small.]	11
II. Deimos		14,500	1 06 18		12

SATELLITES OF JUPITER.

	° ′ ″				
I. Io	3 05 30	268,000	1 18 28	2400	7
II. Europa	3 04 25	426,000	3 13 15	2100	7
III. Ganymede	3 00 28	679,000	7 03 43	3500	6
IV. Callisto	2 58 48	1,200,000	16 16 32	3090	7

SATELLITES OF SATURN.

I. Mimas		122,000	0 22 37	1000	17
II. Enceladus	[The first 7 move very nearly in the plane of the planet's equator: the 8th in a plane inclined about 12° to that plane]	156,000	1 08 53		15
III. Tethys		193,000	1 21 18	500	13
IV. Dione		248,000	2 17 41	500	12
V. Rhea		347,000	4 12 25	1200	10
VI. Titan		805,000	15 22 41	3300	8
VII. Hyperion		1,018,000	21 07 07		17
VIII. Iapetus		2,334,000	79 07 54	1800	9

SATELLITES OF URANUS.

I. Ariel	[The motion of the satellites is retrograde, in a plane inclined 79° to the plane of the ecliptic.]	124,000	2 12 48		
II. Umbriel		173,000	4 03 27		
III. Titania		284,000	8 16 56		
IV. Oberon		380,000	13 11 07		

SATELLITE OF NEPTUNE.

I.	29° to ecliptic; motion retrograde	220,000	5 21 08		14

TABLE V.

THE MINOR PLANETS.

No.	Name.	Year of Discovery.	Discoverer.	Inclination.	Eccentricity.	Period in Years.	Distance in Radii of Earth's orbit.	Diameter in miles.
1	Ceres........	1801	Piazzi..........	10° 36'	0.080	4.6	2.77	227
2	Pallas......	1802	Olbers.........	34 42	0.240	4.6	2.77	172
3	Juno........	1804	Harding......	13 03	0.256	4.4	2.67	112
4	Vesta.......	1807	Olbers........	7 08	0.090	3.6	2.36	228
5	Astræa......	1845	Hencke.......	5 19	0.190	4.1	2.58	61
6	Hebe........	1847	Hencke.......	14 46	0.201	3.8	2.43	100
7	Iris	......	Hind...........	5 27	0.231	3.7	2.39	96
8	Flora.......	......	Hind...........	5 53	0.157	3.3	2.20	60
9	Metis.......	1848	Graham.......	5 36	0.123	3.7	2.39	76
10	Hygeia.....	1849	De Gasparis..	3 47	0.101	5.6	3.15	111
11	Parthenope	1850	De Gasparis .	4 36	0.099	3.8	2.45	62
12	Victoria....	......	Hind...........	8 23	0.219	5.6	2.33	41
13	Egeria......	......	De Gasparis..	16 32	0.088	4.1	2.58	73
14	Irene........	1851	Hind...........	9 07	0.165	4.2	2.59	68
15	Eunomia...	......	De Gasparis..	11 44	0.188	4.3	2.64	12
16	Psyche......	1852	De Gasparis..	3 04	0.136	5.0	2.93	93
17	Thetis	......	Luther.........	5 35	0.127	3.9	2.47	52
18	Melpomene	......	Hind...........	10 09	0.217	3.5	2.30	54
19	Fortuna....	......	Hind...........	1 32	0.158	3.8	2.44	61
20	Massilia....	......	De Gasparis..	0 41	0.144	3.7	2.41	68
21	Lutetia.....	......	Goldschmidt.	3 05	0.162	3.1	2.44	40
22	Calliope....	......	Hind...........	13 44	0.104	5.0	2.91	96
23	Thalia......	......	Hind...........	10 13	0.232	4.3	2.62	42
24	Themis	1853	De Gasparis.	0 48	0.117	5.6	3.14	36
25	Phocea......	......	Chacornac....	21 34	0.253	3.7	2.40	31
26	Proserpine.	......	Luther	3 35	0.088	4.3	2.66	47
27	Euterpe	......	Hind...........	1 35	0.173	3.6	2.35	39
28	Bellona.....	1854	Luther	9 21	0.150	4.6	2.78	59
29	Amphitrite	......	Marth..........	6 07	0.072	4.1	2.55	83
30	Urania......	......	Hind...........	2 05	0.127	3.6	2.36	51
31	Euphrosyne	......	Ferguson	26 25	0.216	5.6	3.16	50
32	Pomona....	......	Goldschmidt.	5 29	0.082	4.2	2.58	35
33	Polyhymnia	......	Chacornac....	1 56	0.338	4.8	2.86	38
34	Circe.......	1855	Chacornac....	5 26	0.110	4.4	2.68	29
35	Leucothea.	......	Luther.........	8 10	0.214	5.2	3.01	25
36	Atalanta ...	......	Goldschmidt.	18 42	0.298	4.6	2.75	20
37	Fides	......	Luther	3 07	0.175	4.3	2.64	41
38	Leda........	1856	Chacornac....	6 58	0.156	4.5	2.74	29
39	Lætitia......	......	Chacornac....	10 21	0.111	4.6	2.77	87
40	Harmonia.	......	Luther	4 15	0.046	3.4	2.27	
41	Daphne....	......	Goldschmidt.	16 45	0.270	4.6	2.77	
42	Isis..........	......	Pogson	8 35	0.226	3.8	2.44	
43	Ariadne....	1857	Pogson	3 27	0.168	3.3	2.20	
44	Nysa........	......	Goldschmidt.	3 41	0.149	3.8	2.42	
45	Eugenia....	......	Goldschmidt.	6 34	0.082	4.5	2.72	
46	Hestia	......	Pogson	2 17	0.162	4.0	2.52	
47	Melete......	......	Goldschmidt.	8 01	0.237	4.2	2.60	

TABLE V.—Continued.

No.	Name.	Year of Discovery.	Discoverer.	Inclination.	Eccentricity.	Period in Years.	Distance in Radii of Earth's orbit.	Diameter in Miles.
48	Aglaia	1857	Luther	5° 00′	0.128	4.9	2.88	
49	Doris		Goldschmidt	6 29	0.076	5.5	3.10	
50	Pales		Goldschmidt	3 08	0.238	5.4	3.09	
51	Virginia		Ferguson	2 47	0.287	4.3	2.65	
52	Nemausa	1858	Laurent	10 14	0.063	3.7	2.38	
53	Europa		Goldschmidt	7 24	0.004	5.5	3.10	
54	Calypso		Luther	5 07	0.213	4.2	2.61	
55	Alexandra		Goldschmidt	11 47	0.199	4.6	2.71	
56	Pandora		Searle	7 20	0.139	4.6	2.77	
57	Mnemosyne	1859	Luther	15 04	0.108	5.6	3.16	
58	Concordia	1860	Luther	5 02	0.041	4.4	2.70	
59	Danäe		Goldschmidt	18 17	0.163	5.1	2.97	
60	Olympia		Chacornac	8 36	0.119	4.5	2.71	
61	Erato		Förster	2 12	0.170	5.5	3.13	
62	Echo		Ferguson	3 34	0.185	3.7	2.39	
63	Ausonia	1861	De Gasparis	5 45	0.127	3.7	2.40	
64	Angelina		Tempel	1 19	0.125	4.4	2.68	
65	Cybele		Tempel	3 28	0.120	6.7	3.42	
66	Maia		Tuttle	3 04	0.154	4.3	2.65	
67	Asia		Pogson	5 59	0.184	3.8	2.42	
68	Hesperia		Schiaparelli	8 28	0.175	5.2	2.99	
69	Leto		Luther	7 58	0.186	4.6	2.77	
70	Panopea		Goldschmidt	11 39	0.183	4.2	2.61	
71	Feronia		C.H.F. Peters	5 24	0.120	3.4	2.27	
72	Niobe		Luther	23 19	0.174	4.6	2.76	
73	Clytie	1862	Tuttle	2 25	0.044	4.3	2.66	
74	Galatea		Tempel	3 59	0.238	4.6	2.78	
75	Eurydice		Peters	5 00	0.307	4.4	2.67	
76	Freia		D'Arrest	2 02	0.188	6.2	3.39	
77	Frigga		Peters	2 28	0.136	4.4	2.67	
78	Diana	1863	Luther	8 39	0.205	4.2	2.62	
79	Eurynome		Watson	4 37	0.195	3.8	2.44	
80	Sappho	1864	Pogson	8 37	0.200	3.5	2.30	
81	Terpsichore		Tempel	7 56	0.212	4.8	2.86	
82	Alcmene		Luther	2 51	0.226	4.6	2.76	
83	Beatrix	1865	De Gasparis	5 02	0.084	3.8	2.43	
84	Clio		Luther	9 22	0.238	3.6	2.37	
85	Io		Peters	11 56	0.194	4.3	2.66	
86	Semele	1866	Tietjen	4 48	0.205	5.4	3.09	
87	Sylvia		Pogson		0.081	6.5	3.49	
88	Thisbe		Peters	5 09	0.167	4.6	2.75	
89	Julia		Stéphan	15 13	0.205	4.0	2.53	
90	Antiope		Luther		0.148	5.5	31.2	
91	Ægina		Stéphan					

The number of minor planets discovered up to October, 1878, was 192. The diameters are derived from photometric experiments made by Professor Stampfer, of Vienna. They are probably only relatively correct.

TABLE VI.

SCHWABE'S OBSERVATIONS OF THE SOLAR SPOTS.

Year.	Number of days of observation.	New Groups.	Days on which the Sun was free from spots.	Year.	Number of days of observation.	New Groups.	Days on which the Sun was free from spots.
1826	277	118	22	1846	314	157	1
7	273	161	2	7	276	257	0
8	282	225(Max.)	0	8	278	330(Max.)	0
9	244	199	0	9	285	238	0
1830	217	190	1	1850	308	186	2
1	239	149	3	1	308	151	0
2	270	84	49	2	337	125	2
3	247	33(Min.)	139	3	299	91	3
4	273	51	120	4	334	67	65
5	244	173	18	5	313	79	146
6	200	272	0	6	321	34(Min.)	193
7	168	333(Max.)	0	7	324	98	52
8	202	282	0	8	335	188	0
9	205	162	0	9	343	205	0
1840	263	152	3	1860	332	211(Max.)	0
1	283	102	15	1	322	204	0
2	307	68	64	2	317	160	3
3	312	34(Min.)	149	3	330	124	2
4	321	52	111	4	325	130	4
1845	332	114	29	1865	307	93	25

TABLE VII.

PERIODIC COMETS.

Name.	Last perihelion passage observed.	Longitude of Asc. Node	Inclination.	Eccentricity.	Semi-major axis.	Period in Days.	Motion.
Encke's	1878	334 31	13 05	0.847	2.22	1210	Direct.
Winnecke's	1875, Mar. 12	113 31	10 48	0.755	3.14	2020	Direct.
Brorsen's	1879, Oct. 12	101 46	29 45	0.802	3.14	2037	Direct.
Biela's	1852, Sept. 23	245 57	12 34	0.756	3.50	2415	Direct.
D'Arrest's	1877	148 27	13 56	0.660	3.44	2366	Direct.
Faye's	1873, July 18	209 40	11 22	0.558	3.81	2715	Direct.
Méchain's	1871, Dec. 1	268 54	54 32	0.830	6.03	4986	Direct.
Halley's	1835, Nov. 16	55 10	17 45	0.967	17.99	28000	Retro.

TABLE VIII.

TRANSITS OF THE INFERIOR PLANETS.

Mercury.		Venus.	
1802.	November 8.	1639.	December 4.
1815.	November 11.	1761.	June 5.
1822.	November 4.	1769.	June 3.
1832.	May 5.	1874.	December 8.
1835.	November 7.	1882.	December 6.
1845.	May 8.	2004.	June 7.
1848.	November 9.	2012.	June 5.
1861.	November 11.	2117.	December 10.
1868.	November 4.	2125.	December 8.
1878.	May 6.	2247.	June 11.
1881.	November 7.	2255.	June 8.
1891.	May 9.	2360.	December 12.
1894.	November 10.	2368.	December 10.

TABLE IX.

LIST OF STARS WHOSE ANNUAL PARALLAX HAS BEEN COMPUTED.

(All these are doubtful.)

Stars.	Parallax.	Distance in radii of the Earth's orbit.	Observer.
α Centauri	0.92″	224,000	Maclear. Henderson.
61 Cygni	{ 0.35 { 0.56	589,000 368,000	Bessel. Auvers.
21258 Lalande	0.27	764,000	Auvers.
17415 Oeltzen	0.25	825,000	Krüger.
1830 Groombridge	0.23	897,000	C. A. Peters.
70 Ophiuchi	0.16	1,289,000	Krüger.
α Lyræ	0.155	1,331,000	W. & O. Struve.
Sirius	0.150*	1,375,000	Henderson. Peters.
ι Ursæ Majoris	0.133	1,550,000	C. A. Peters.
Arcturus	0.127	1,624,000	C. A. Peters.
Polaris	0.07	2,950,000	C. A. Peters.
Capella	0.05	4,130,000	C. A. Peters.

* Another determination gives 0″.23.

TABLE X.

THE PRINCIPAL CONSTELLATIONS.

Those found in Ptolemy's Catalogue (137 A.D.) are in *Italics*.

- THE NORTHERN CONSTELLATIONS.

No.	Name.	Coordinates of Centre.			Name of Principal Star of 1st or 2d magnitude.	Number of stars of 1st mag.	Number of stars of first five magnitudes.
		R.A.		D.			
		h.	m.	°			
1	*Andromeda*	1	0	35	Alpheratz (a)	1	18
2	*Cassiopeia*	1	10	60			46
3	*Triangulum*	2	0	32			5
4	*Perseus*	3	30	47	Mirfak (a)		40
5	Camelopardus	5	45	68			36
6	*Auriga*	6	0	42	Capella (a)	1	35
7	Lynx	7	55	50			28
8	Leo Minor	10	05	36			15
9	*Ursa Major*	10	40	58	Dubhe (a)	1	53
10	Coma Berenicis	12	40	26			20
11	Canes Venatici	13	0	40	Cor Caroli (a)		15
12	*Boötes*	14	35	30	Arcturus (a)	1	35
13	*Ursa Minor*	15	0	78	Polaris (a)		23
14	*Corona Borealis*	15	40	30			19
15	*Serpens*	15	40	10	Unukalhay (a)		23
16	*Hercules*	16	45	27	Rasalgeti (a)		65
17	*Draco*	17	20	66	Thuban (a)		80
18	TaurusPoniatowskii	17	50	5			6
19	Clypeus Sobieskii	18	10	15 S.			4
20	*Lyra*	18	40	35	Vega (a)	1	18
21	*Aquila*	19	30	10	Altair (a)	1	33
22	*Sagitta*	19	40	18			5
23	Vulpecula et Anser.	20	0	25			23
24	*Cygnus*	20	20	42	Deneb (a)		67
25	Lacerta	20	20	44			13
26	*Delphinus*	20	40	15			10
27	*Equuleus*	21	0	6			5
28	*Cepheus*	21	40	65			44
29	*Pegasus*	22	25	15	Markab (a)		43
29	Total					6	827

TABLE X.—*Continued.*

THE ZODIACAL CONSTELLATIONS.

No.	Name.	Coordinates of Centre. R.A. h. m.	Coordinates of Centre. D.	Name of Principal Star of 1st or 2d magnitude.	Number of stars of 1st mag.	Number of stars of first five magnitudes.
1	*Aries*	2 30	18 N.	Hamal (*a*)		17
2	*Taurus*	4 0	18	Aldebaran (*a*)	1	58
3	*Gemini.*	7 0	25	{ Castor (*a*) { Pollux (*β*)	1	28
4	*Cancer*	8 40	20			15
5	*Leo*	10 20	15	Regulus (*a*)	1	47
6	*Virgo*	13 20	3 N.	Spica (*a*)	1	39
7	*Libra*	15 0	15 S.	Zubenelg (*a*)		23
8	*Scorpio*	16 15	26	Antares (*a*)	1	34
9	*Sagittarius*	18 55	32			38
10	*Capricornus*	21 0	20	SecundaGiedi(*a*)		22
11	*Aquarius*	22 0	9 S.	Sadalmelik (*a*)		25
12	*Pisces*	0 20	10 N.			18
12	Total				5	364

THE SOUTHERN CONSTELLATIONS.

1	Apparatus Sculptoris.	0 20	32			13
2	Phœnix	1 0	50			32
3	*Cetus*	2 0	12	Menkar (*a*)		32
4	Fornax Chemica	2 20	30			6
5	Hydrus	2 40	70			25
6	Horologium	3 15	57			11
7	*Eridanus*	3 40	30	Achernar (*a*)	1	64
8	Reticulus Rhomboidalis	4 0	62			11
9	Dorado	4 40	62			17
10	Cæla Sculptoris	4 40	42			6
11	Mons Mensæ	5 20	75			9
12	Columba Noachi	5 25	35	Phact (*a*)		15
13	Equuleus Pictoris	5 25	55			17
14	*Lepus*	5 25	20	Arneb (*a*)		18
15	*Orion*	5 30	0	Rigel (*β*)	2	37
16	Canis Major	6 45	24	Sirius (*a*)	1	27
17	Monoceros	7 0	2			12
18	*Canis Minor*	7 25	5	Procyon (*a*)	1	6
19	*Argo*	7 40	50	Canopus (*a*)	2	133
20	Piscis Volans	7 40	68			9
21	Sextans	10 0	0			3

TABLES. 265

TABLE X.—*Continued.*

THE SOUTHERN CONSTELLATIONS.

No	Name.	Coordinates of Centre. R.A. h. m.		D. °	Name of Principal Star of 1st or 2d magnitude.	Number of stars of 1st mag.	Number of stars of first five magnitudes.
22	*Hydra*	10	0	10	Alphard (a)		49
23	Antlia Pneumatica	10	0	35			7
24	Chamæleon	10	50	78			17
25	*Crater*	11	20	15			9
26	Crux	12	15	60		1	10
27	*Corvus*	12	20	18	Algorab (a)		8
28	Musca Australis	12	25	68			7
29	*Centaurus*	13	0	48		2	54
30	Circinus	15	0	64			2
31	Apus	15	20	76			7
32	*Lupus*	15	25	45			34
33	Triangulum Australe	15	40	65			11
34	Norma	16	0	45			12
35	*Ara*	17	0	54			15
36	Ophiuchus	17	0	0	Rasalague (a)		40
37	*Corona Australis*	18	30	40			7
38	Telescopium	18	40	53			8
39	Pavo	19	20	68			27
40	Microscopium	20	40	37			5
41	Indus	21	0	55			15
42	Octans	21	0	80			16
43	*Piscis Australis*	21	40	32	Fomalhaut (a)	1	16
44	Grus	22	20	47			11
45	Toucan	23	45	66			21
45	Total					11	911
86	Summary					22	2102

TABLE XI.

VARIABLE STARS.

Name.	Coordinates, 1870. R.A.	D.	Period in Days.	Change of Magnitude.	Authority.	
	h. m. s.	° ′				
α Cassiopeiæ.	0 33 09	55 49.4 N	79.1	2 to 2.5	Birt,	1831
ο Ceti..........	2 12 47	3 34.1 S	331.34	2 " 12	D. Fabricius,	1596
β Persei.......	2 59 43	40 27.2 N	2.87	2.5 " 4	Montanari,	1669
λ Tauri........	3 53 29	12 07.3 N	3.95	4 " 4.5	Baxendell,	1848
ε Aurigæ.....	4 52 38	43 37.7 N	250.	3.5 " 4.5	Heis,	1846
μ Doradûs....	5 05 47	61 58.4 S	Long.	5 " 9	Moesta,	1865
α Orionis......	5 48 08	7 22.8 N	196.	1 " 1.5	J. Herschel,	1836
ζ Geminorum	6 56 24	20 45.5 N	10.16	3.8 " 4.5	Schmidt,	1847
α Hydræ......	9 21 12	8 05.9 S	55.	2.5 " 3	J. Herschel,	1837
R Leonis......	9 40 34	12 01.8 N	331.	5 " 11.5	Koch,	1782
η Argûs........	10 40 02	59 0.1 S	46 yrs.	1 " 4	Burchell,	1827
R Hydræ......	13 22 37	22 36.4 S	447.8	4 " 10	Maraldi,	1704
Z Virginis....	13 27 45	12 32.7 S	?	5 " 8	Schmidt,	1866
T Coronæ.....	15 54 04	26 17.5 N	?	2.5 " 9.8	Bermingham,	1866
30 Herculis...	16 24 22	42 10.1 N	106.	5 " 6	Baxendell,	1857
Nov. Ophiu...	16 52 13	12 41.4 S	?	4.5 " 13.5	Hind,	1848
α Herculis....	17 08 43	14 32.4 N	?	3.1 " 3.9	W. Herschel,	1795
R Clyp. Sob.	18 40 33	5 50.5 S	89.	5 " 9	Pigott,	1795
β Lyræ.......	18 45 17	33 12.7 N	12.91	3.5 " 4.5	Goodricke,	1784
13 Lyræ.......	18 51 23	43 46.6 N	46.	4.2 " 4.6	Baxendell,	1855
χ² Cygni......	19 45 34	32 35.2 N	409.2	5 " 13	Kirch,	1686
η Aquilæ......	19 45 51	0 40.4 N	7.18	3.6 " 4.4	Pigott,	1784
34 Cygni......	20 13 0	37 37.8 N	18 yrs.	3 " 6	Jansen,	1600
R Cephei......	20 23 41	88 44. N	73 yrs.	5 " 11	Pogson,	1850
μ Cephei......	21 39 31	58 11.1 N	5 or 6 yrs.	4 " 6	W. Herschel,	1782
δ Cephei......	22 24 21	57 45. N	5.37	3.7 " 4.8	Goodricke,	1784
β Pegasi......	22 57 28	27 22.7 N	?	2 " 2.5	Schmidt,	1848

TABLE XII.

BINARY STARS.

Name.	Co-ordinates, 1870. R.A.	D.	Magnitude of Components.	Semi-major Axis.	Eccentricity.	Period in Years.	Calculator.
ζ Herculis....	h. m. s. 16 36	° ′ 31 51 N	3 — 6	″ 1.25	0.45	36.3	Villarceau.
η CoronæBor..	15 18	30 47 N	6 — 6½	0.95	0.28	43.6	Winnecke.
ζ Cancri.......	8 04 45	18 02.4 N	6–7–7½	1.29	0.23	58.9	Mädler.
ξ Ursæ Maj...	11 11 15	32 16 N	4 — 5½	2.45	0.39	63.1	J. Breen.
α Centauri....	14 30 48	60 17.9 S	1 — 2	15.50	0.95	80.0	E.B.Powell.
ω Leonis......	9 21	9 39 N	6½ — 7½	0.85	0.64	82.5	Mädler.
τ Ophiuchi...	17 56	8 11 S	5 — 6	0.82	0.037	87.	Mädler.
70 Ophiuchi..	17 58 52	2 22.5 N	4½ — 7	4.19	0.44	92.8	Mädler.
λ Ophiuchi...	16 24	2 17 N	4 — 6	0.84	0.477	95.	Hind.
γ CoronæAus.	18 57 38	37 14.7 S	6 — 6	2.54	0.60	100.8	Jacob.
ξ Boötis........	14 45 22	19 38.5 N	3½ — 6½	12.56	0.59	117.1	J.Herschel.
δ Cygni.......	19 41	44 48 N	3½ — 9	1.81	0.60	178.7	Hind.
η Cassiopeiæ.	0 41 09	57 07.8 N	4 — 7½	10.33	0.77	181.	E.B.Powell.
γ Virginis....	12 35 05	0 44.2 S	4 — 4	3.58	0.87	182.1	J.Herschel.
σ CoronæBor.	16 10	34 12 N	6 — 6½	2.71	0.30	195.1	Jacob.
Castor	7 26 18	32 10.4 N	3 — 3½	8.08	0.75	252.6	J.Herschel.
61 Cygni......	21 01 04	38 05.1 N	5½ — 6	15.4		452.	
μ Boötis........	15 19 35	37 50 N	4 — 8	3.21	0.84	649.7	Hind.
γ Leonis	10 12 47	20 30.1 N	2 — 4			1200.	

INDEX.

[The references are to the pages.]

Aberration of light, 116; diurnal and annual, 117; separated from parallax, 221.
Acceleration, secular of the moon, 133.
Aerolites, 209.
Algol, 224.
Altazimuth, 39.
Altitude, 19.
Altitude and azimuth instrument, 38; use of, 40.
Altitudes, method of equal, 39.
Amplitude, 19.
Andromeda, nebula in, 231.
Annular eclipse, 141.
Anomaly, 90.
Aphelion, 90.
Apogee, 122.
Appulse, 136.
Apsides, of earth, 116; of moon, 122; of planets, 164.
Arc of meridian measured, 59.
Argo, nebula in, 234.
Aries, first point of, 20, 84.
Ascension, right, 20; related to sidereal time, 22.
Asteroids, 171; table of, 260.
Astronomy, 11; chronology of, 248.
Atmosphere, height of, 52.
Attraction, law of, 108.
Axis, of the heavens, 15; of the earth, 17; of rotation and collimation, 32.
Azimuth, 19.

Base-line, 60.
Bode's law, 171.
Branches of meridian, 18.

Calendar, 105.
Centauri, alpha, distance of, 222; a binary star, 227.
Centrifugal force, 65, 247.
Ceres, discovery of, 172.

Chronograph, 29.
Chronometer, Greenwich time given by, 77.
Circle, vertical, 19; hour, 20; of perpetual apparition, 20; diurnal, 24; of latitude, 84.
Circle, meridian, 34; mural, 38; reflecting, 48.
Clock, astronomical, 28; driving, 41.
Clusters of stars, 229.
Coal sack, 234.
Collimation, axis of, 32.
Colures, 84.
Comets, 187; diversity of appearance, 188; tail, 189; orbits, 191; periods and motion, 192; mass and density, 193; light, 194; periodic, 195; Encke's, 195; Winnecke's or Pons's, 197; Brorsen's, 197; Biela's, 197; D'Arrest's, 198; Faye's, 198; Méchain's, 199; Halley's, 199; Great, of 1811, 200; of 1843, 200; Donati's, 201; of 1861, 202; connection with meteors, 211; elements of periodic, 261.
Compression, 63.
Conjunction, 126; inferior and superior, 159.
Constellations, 215; list of, 263.
Co-ordinates, spherical, 25.
Corona, 96.
Count, least, 47.
Crescent, 127.
Cross-wires, 32.
Culmination, 20.
Cycle, lunar, 133; of eclipses, 142.
Cygni, 61, its distance, 222.

Day, solar and sidereal, 21; inequality of solar, 91; astronomical and civil, 105; intercalary, 106.
Declination, 20.
Degree of meridian, 62.

Departure, 18.
Dip of the horizon, 57.
Dipper, 215.
Disc, spurious, of stars, 223.
Distance, zenith, 19; polar, 20.

Earth, general form of, 12; spheroidal form, 62; dimensions, 63; density, 63; linear velocity of rotation, 70; revolution about the sun, 88; orbital velocity, 89; orbit, 89; motion at perihelion and aphelion, 110; revolution proved by aberration, 119; phases, 127; elements, 257.
Eccentricity of an ellipse, 90.
Eclipses, 135; lunar, 135; solar, 139; total, 142; cycle of, 142; number of, 143; of Jupiter's satellites, 174.
Ecliptic, 83.
Ecliptic limits, lunar, 137; solar, 141.
Elements, of planetary orbit, 160; of cometary, 190.
Ellipse, 244.
Elongation, 56; greatest eastern and western, 159.
Equation, of centre, 102; of time, 104; annual, 133.
Equator, 17; celestial, 19.
Equatorial, 40.
Equilibrium of centrifugal and centripetal forces, 107.
Equinoctial, 19.
Equinox, vernal, 20, 83.
Error of clock, 28.
Errors of observation, 49.
Establishment of port, 151.
Evection, 132.
Evening star, 160, 170.

Faculæ, 96.
Finder, 34.
Flames, red, 97.
Forces, centrifugal and centripetal, 247.
Foucault's experiment, 68.

Galaxy, 234.
Gemini, 216.
Geocentric, parallax, 55; motion of planets, 155, 158, 167.

Gibbous, 127.
Golden number, 134.
Gravitation, universal, 107.
Heliocentric, parallax, 56, 156; motion of planets, 157.
Hemisphere, 17.
Horizon, 13; points of, 19; artificial, 45.
Horizontal point, 36.
Hour angle, 20; relation to sidereal time, 22.
Hyades, 230.
Hyperbola, 245.

Incidence, angle of, 51.
Index correction, 44.

Jupiter, 173; mass of, 175.

Kepler's laws, 111.
Kirkwood's law, 247.

Latitude, 18; equal to altitude of pole, 23; methods of determining, 72; at sea, 75; reduction of, 75; celestial, 84.
Level, hanging, 34.
Librations, 131.
Light, analysis of, 48; of sun, 97; velocity of, 118; of planets, 183; of stars, 218; of nebulæ, 230.
Line of sight, 32.
Longitude, 18; how determined, 76; by telegraph, 78; by star signals, 79; at sea, 80; celestial, 84; by eclipses and occultations, 145.
Luculi, 96.
Lunar distance, 78.
Lunation, 128.

Magellanic clouds, 233.
Magnitudes, 214.
Mars, 170.
Mercury, 164; transits of, 262.
Meridian, 18; prime and celestial, 18; line, 19.
Meteors, 202; showers, 203; height and velocity, 205; orbits, 206; detonating, 208.
Microscope, reading, 35.

INDEX. 271

Milky way, 234.
Minor planets, 171; list of, 260.
Mira, or *o* Ceti, 223.
Moon, orbit of, 120; nodes, 120; obliquity of orbit, 121; form of orbit, 122; line of apsides, 122; meridian zenith distance, 122; distance, 123; magnitude and mass, 125; augmentation of diameter, 125; phases, 126; sidereal and synodic periods, 128; retardation, 129; harvest, 130; rotation, 130; librations, 131; other perturbations, 132; general description, 134; elements, 257.
Morning star, 160, 170.
Motion, diurnal, 13; upward and downward, 18; west to east, 88 (note); direct and retrograde, 159, 167.
Mural circle, 38.

Nadir, 18; point, 37.
Navigation, sketch of, 255.
Nebulæ, resolvable and irresolvable, 230; annular and elliptic, 231; spiral and planetary, 232; nebulous stars, 232; double, 232; variation of brightness, 233.
Nebular hypothesis, 183.
Neptune, 182; immense distance of, 183.
Nodes, of moon's orbit, 120; heliocentric longitude of planet's, 161.
Noon, 105.
Nubeculæ, 233.
Nutation, 115.

Obliquity of ecliptic, 84, 115.
Occultation, 135, 144; of Jupiter's satellites, 174.
Octant, 48.
Olbers's theory, 172.
Opposition, 126.
Orion, 216.

Parabola, 246.
Parallax, geocentric and horizontal, 55; heliocentric. 56, 156; annual, 219.
Pegasus, 216

Pendulum experiment, 68.
Penumbra, of solar spots, 95; of eclipse, 136.
Perigee, 122.
Perihelion, 90.
Perturbations, in earth's orbit, 112; in moon's, 130.
Phases, of moon, 126; of earth, 127; of Mercury and Venus, 165; of Mars, 171.
Photosphere, 95.
Planetoids, 171; list of, 260.
Planets, 155; orbits of, 157; inferior, 158; stationary points, 159, 168; elements of orbit, 160; heliocentric longitude of node, 161; inclination of orbit, 162; periods, 163, 168; superior, 167; distance, 169; elements, 256.
Plateau's experiment, 185.
Pleiades, 230.
Points, fixed, 36.
Pointers, 216.
Poles, of the heavens, 15, 17; of the earth, 17.
Pole-star, 14.
Position angle, 24.
Præsepe, 230.
Precession, 112.
Problem of three bodies, 133.
Projections, spherical, 27.
Proper motions, 213.

Quadrant, 48.
Quadrature, 126.

Radiant points, 206.
Rate of clock, 28.
Refraction, 51; astronomical, 52; general laws, 53; effects of, 54.
Resisting medium, 196.
Reticule, 32.
Retrogradation, 159, 167.
Rings of Saturn, 177; disappearance of, 178.

Saros, 143.
Satellites, elements of, 258.
Saturn, 176; rings of, 177.
Seasons, 90.

Sextant, 42; prismatic, 48.
Shadow, of earth, 136; of moon, 139.
Signs of zodiac, 84.
Sirius, light of, 223; orbit of, 240.
Solar system, 11; orbit of, 238.
Solstices, 84.
Spectroscope, 48; use of, 97.
Sphere, celestial, 11; parallel, 16; right and oblique, 17.
Spheroid, oblate, 63, 245.
Spots, solar, 95; observations of, 261.
Star signals, 79.
Stars, circumpolar, 14; fixed, 213; number of, 214, 234; magnitudes, 214; of first magnitude, 217; constitution, 218; distance, 218; differential observations, 220; real magnitudes, 222; variable and temporary, 223; double and binary, 225; colored, 228; examples of variable, 266; of binary, 267.
Stationary points, 159, 168.
Style, old and new, 106.
Sun, distance of, 84; magnitude, 87; rotation, 95; constitution, 98; irregular advance, 102; first mean, 102; second mean, 103; mass and density, 109; size compared with stars, 223; motion in space, 235; elements, 256.
Synodical revolution, of moon, 128; of planets, 163, 168.

Talcott's method of finding latitude, 74.
Telegraph, used in determining longitude, 78.
Telescopic comets, 188.
Tempel's comet, leads November shower, 212.
Theodolite, 48.
Tides, 146; daily inequality, 148; general laws, 149; influence of sun, 149; spring and neap, 150; priming and lagging, 150; tidal wave, 151; establishment, 151; cotidal lines, 152; height, 152; four daily, 153; in lakes, 154.
Time, solar and sidereal, 21; sidereal and right ascension, 22; Greenwich, 77; local time at different meridians, 81; astronomical and civil, 105.
Torsion balance, 63.
Trade-winds, 67.
Transit instrument, 31.
Transit, 20; of inferior planets, 262.
Triangle, astronomical, 23.
Triangulation, 60.
Twilight, 94.

Umbra, of solar spots, 95; of eclipses, 136.
Universal instrument, 48.
Uranus, 180.
Ursa, major, 215; minor, 216.

Vanishing points and circles, 26.
Variation, 133.
Venus, relative distances from sun and earth, 84; transit of, 86, 165, 262; description of, 165.
Vernier, 46.
Vertical, lines, 18; prime, 19.
Vulcan, 157 (note).

Weight, in different latitudes, 64; on the sun, 110.

Year, sidereal, 83; tropical, 105, 114; anomalistic, 116.

Zenith, 18; geographical and geocentric, 76.
Zenith telescope, 48; use of, 74.
Zodiac, 84.
Zodiacal light, 99.

THE END.

www.ingramcontent.com/pod-product-compliance
Lightning Source LLC
Chambersburg PA
CBHW031936230426
43672CB00010B/1945